The Digital and Its Discontents

Electronic Mediations

Series Editors: N. Katherine Hayles, Peter Krapp,
Rita Raley, and Samuel Weber

Founding Editor: Mark Poster

(continued on page 245)

The Digital and Its Discontents

Aden Evens

Foreword by Alexander R. Galloway

Electronic Mediations 62

University of Minnesota Press
Minneapolis
London

The University of Minnesota Press gratefully acknowledges the financial assistance provided for the publication of this book by Dartmouth College.

Portions of chapter 3 were previously published in "Digital Ontology and Contingency," in *Contingency and Plasticity in Everyday Technologies,* ed. Natasha Lushetich, Iain Campbell, and Dominic Smith, 35–52 (London: Rowman & Littlefield, 2022); all rights reserved.

Published by the University of Minnesota Press
111 Third Avenue South, Suite 290
Minneapolis, MN 55401-2520
http://www.upress.umn.edu

ISBN 978-1-5179-1631-2 (hc)
ISBN 978-1-5179-1632-9 (pb)

A Cataloging-in-Publication record for this book is available from the Library of Congress.

Printed on acid-free paper

The University of Minnesota is an equal-opportunity educator and employer.

For my younger son, Jules.
No doubt he'll write his own book before too long.

. . . und die Zahl gerade ist das Erstaunlichste in den Dingen.
—Friedrich Nietzsche,
"Über Wahrheit und Lüge im außermoralischen Sinne"

Imperfection, ambiguity, opacity, disorder, and the opportunity
to err, to sin, to do the wrong thing: all of these are constitutive of
human freedom, and any concentrated attempt to root them out
will root out that freedom as well. If we don't find the strength
and the courage to escape the silicon mentality that fuels much
of the current quest for technological perfection, we risk finding
ourselves with a politics devoid of everything that makes politics
desirable, with humans who have lost their basic capacity for
moral reasoning, with lackluster (if not moribund) cultural
institutions that don't take risks and only care about their financial
bottom lines, and, most terrifyingly, with a perfectly controlled
social environment that would make dissent not just impossible
but possibly even unthinkable.
—Evgeny Morozov, *To Save Everything, Click Here*

Contents

Foreword

Alexander R. Galloway

Some books on digital technology are descriptive. These books will often begin from digital culture or from a user's phenomenological experience with machines, in order to push a synthetic claim benefiting from that vantage. These books might build intellectual momentum from archival or empirical research, trying to uncover technical artifacts lost to history, or to rearticulate the meaning of technical devices that saturate such a history.

Other books focus more on definitions and axioms. They plumb the foundations of things. Instead of hunting the pertinent detail, they prize the general convention, sometimes even the universal law. Taking a cue from the rationalists more than the empiricists, these kinds of books build their cases by defining the conditions of possibility for anything whatsoever. Instead of specific details anticipating general claims, here the general stipulates a framework within which all sorts of specifics might emerge.

The Digital and Its Discontents is a book about the messy ambiguity of earthly things, to borrow an appealing expression from the author. The book is also about the banishment of such messy contingency through the irrepressible necessity of digital logic. Contingency and necessity—that discontented pair, locked in a discontented union—generate the discontents of the book's title. And, while finely attuned to the world around us, this book is not primarily focused on the culture or politics of computation. As Aden Evens himself puts it, this is "a work of philosophy rather than cultural theory." The book offers a critique of the digital overall, by looking at the general principles that underlie digital computation.

Even today, with fiber-optic cables garlanding the planet and digital services managing the population, there are surprisingly few books that theorize the digital explicitly. What is a byte, do we even yet know? What is the philosophical significance of a logical operator? Where do computers fit within the history of thinking? In an earlier time, we might have called this a "structuralist" theory of the digital, or at least an approach that defines the digital in terms of its core structures or foundational formal arrangements. I applaud Evens for it, and if that particular label isn't apt, maybe "rationalist" is, or at the very least "philosophical," for this is a (rationalist) philosophy of the digital more than an empirical description of it.

Okay, so what is the digital exactly? Treading carefully and deliberately, Evens returns the reader to one of the oldest problems in philosophy, the problem of the discrete and the continuous. From the Pythagoreans, to Zeno of Elea, to Euclid, up to G. W. Leibniz and Georg Cantor in the modern era and beyond, many have wrestled with the problem of the discrete and the continuous. Or to put the problem in crass terms: Are there two things, or is there really just one? And, anyway, how do you get two things if you only start with one?

The easiest way to get two things from one thing is to cut it in half. Although beware, because we've already cheated, or at least smuggled in something new: an operation adjoined to a substance. The operation doesn't necessarily have to be defined as "cutting." We might also talk about "distinction" or "differentiation" or "making discrete." In any case, whether as cut, difference, distinction, or discretization, this operation results in some kind of rudimentary digital difference. I'll stress, as Evens does, that this kind of difference needs to be quantitative, which is to say discrete. (There are also ways to think about difference that remain strictly qualitative, yet modern computers are not built around qualitative difference.)

Rudimentary digital difference has a more common name in computation. It's called a *bit*. And Evens grounds his investigation, I think rightly, around these sorts of binary values, commonly expressed in the form of a 1 or a 0, or equally as true/false, on/off, high voltage / low voltage, p/q, or any other pair of terms that can be exactly and uniquely discretized. (In a brilliant moment at the start of chapter 4, Evens all but asserts that the bit is a literal embodiment of Aristotle's "law of excluded middle," a seeming anachronism that, I hope, might fruitfully reshuffle the historical timeline of digital philosophy.) Yet all this cutting is something of a sleight of hand. We might have multiple bits, but not really. Chopping the world up into bits (or atoms, or individuals), into what Euclid defined as arithmetical multiples, in fact produces a new kind of continuousness, a new monotony of the monad. And hence I suspect we can rightly describe digital computation as a technology based on one thing rather than two, despite the profusion of digital difference at its core. Computers are, in other words, a *monist* technology. And the digital difference at the heart of computation is not an exception to such consistency, but precisely its very substance.

The skeptic might interject, "Yes, but bits are two things, zeros and ones, you just said so!" This is true in a narrow sense, yet it misses the larger point. Zero and one are essentially the same type of thing, a thing defined exclusively as having a discretely differentiable value. So it would be a mistake to think that computers institute a binary regime or a regime of duality. Like the integers, binary bits are all one kind of thing, even if different bits represent different values. Computers are a species of monism. And that has made all the difference.

A constitutional problem thus haunts the invention of digital computers, in that *they were never endowed with duality.* Computers begin from one kind of thing, the bit, and don't really have a reliable way to derive other kinds of things. One might say that computers have a technology for *difference* but they don't have a technology for *the concept.* This presents something of a problem when it comes to traditional philosophy. The standard model of metaphysics, for instance, relies on two things, not one, two dramatically different kinds of things, as different as mind and body are different, or as different as God and man are different. And so the hard problem of computation has been, roughly, how to construct two kinds of things from having only one at the outset.

In other words, if René Descartes had a mind–body problem, computers have a body–body problem. Computers lost their minds because they never had them in the first place. There is no ghost in this machine, alas. Engineers have struggled to add the mind somehow, if only in rudiment, by superimposing conceptual differences—we could even say qualitative differences, or a reasonable simulation thereof—onto sequences of bits. These include the basic conceptual difference between two kinds of bits, those that are interpreted as data and those that are interpreted as procedures. *Don't mix these two up!* Or arrange clusters of bits together according to *type*, a metaphysical concept if there ever was one. Or, as Evens specifies, group bits together in spatial and logical arrangements, endowing these arrangements with address indexes, and then allow numbers to refer to other numbers according to these addresses. (In a computer language like C, these references are called "pointers.") This last example is ingenious, because it shifts the computer's monism from bug to feature: a number might simply represent itself, its own value, or it might be serving as a reference to an address, its own value now miraculously interpreted as an arrow pointing to something else entirely. Ingenious, such references are also hazardous, given the learning curve involved in effectively programming using pointers, not to mention the host of potential bugs that tend to accompany them.

But what is this all for? What foundational lack is the digital machine so worried about filling? Here Evens steadfastly resists epistemological questions, which are certainly the more popular option for digital theory thus far, concerned as it has been with representation and simulation, or what we can know using these machines, or even whether these machines can know anything on their own. He turns instead to ontology, or the forms of being and presence wrought by the digital. Not that epistemological discussions are unimportant; we simply already have so many examples of them.

In order to delineate digitality's foundational lack, Evens paints a picture of the world in stark relief. We've heard about the digital side already. The other side is what Evens calls "the actual." If the digital is ruled by the logic of necessity, the actual follows the logic of contingency. If the digital is tangled up in what Evens describes as "a knot of Enlightenment thought," an unholy trinity of positivism, rationalism, and instrumentalism, then the actual accommodates ambiguity, indeterminacy, and inefficiency. Evens furnishes the reader with a series of evocative keywords to help understand the actual: "difference, becoming, freedom, uncertainty, accident, noise, virtuality, groundlessness, faith." At one brief moment, Evens even connects the actual with Jacques Lacan's notion of the "real," which we will take as a suggestive prompt for future investigation. Pushing his vocabulary in creative directions, Evens describes the actual as having a kind of "meshiness," a word stemming from "mesh" or "network" but also containing hints of "messy" or "messiness." Or, in another nice turn of phrase, Evens highlights the rule-bound quality of the digital, dubbed "ruliness," in order to differentiate it more effectively from the characteristic "unruliness" of the actual.

Even the most impressive developments in artificial intelligence do not ultimately disrupt this fundamental arrangement, as Evens discusses toward the end of the book. AI evangelists often point to the difference between AI's "symbolic" phase in the 1950s and 1960s, a confirmed failure, and today's "empiricist" phase, judged by many to be a runaway success. After it became clear that symbolic logic alone would not produce convincing results, AI researchers supplemented their digital methods with a series of analog ones, focusing on empirical data, using inductive logic, embracing stochasticism and probability, and adopting a variety of "living" methods like neural networks, evolutionary algorithms, and cellular automata. Much of this was helped along by improvements in graph theory and clustering algorithms. The results are no doubt spectacular. Yet I agree with Evens's critique that "the concatenation of billions or trillions of trivially simple calculations" doesn't somehow undo or invert the constitutive characteristics of the digital. In effect, digital computers are forever trapped in what G. W. F. Hegel called "bad infinity." Billions or trillions—make it much larger, it doesn't matter. "Sheer numeracy is one of the principal strategies of the digital, as it can make up for a lot of weaknesses," Evens admits at one point. Nevertheless, sheer numeracy will never solve the body–body problem. No mere quantitative extension will ever generate a dialectical inversion, at least not in the Hegelian sense, to say nothing of the Marxist sense. And, in any case, the fact that AI engineers have basically given up on deriving value from discrete rationality alone, in favor of extracting value from empirical data, confirms Evens's conclusions all the more. If value comes out of the black box, this value was no doubt originally sourced from the actual, not the digital.

Those of different philosophical persuasions might quibble over the details, but the core of Evens's metaphysics has been foundational to Western philosophy for a very long time: nature is rich and full of contingency, while machines are rote and deterministic. If modern life is plagued by the discontents of alienation and ennui (or worse), we would certainly benefit from scrutinizing the ontological imbalance on which it rests. "A different outlook" might emerge, he writes. An outlook that would better value "the social over the individual, the continuous over the discrete, becoming over being, and finally difference over identity."

I will leave the reader to discover the many other riches of the book. These include Evens's proposals for a "Principle of Abundant Reason" and a "Law of Digital Hermetics." The brilliant chapter 4 pushes the case to its most profound conclusions. A level of technical precision guides the book throughout, without overwhelming the reader in arcana. And, last but not least, I appreciate Evens's continuous attention to value and meaning, dare I say also to our common humanity, a welcome balm in these times of discontent.

Introduction

It's not just that digital technologies are everywhere. It's that they are so deeply entangled in so much of human life. And we keep finding more ways to put them to use, more opportunities and more reasons to "go digital." This book examines the consequences of conducting so much of our lives in and through the digital, asking how those technologies influence our behaviors, our ways of understanding ourselves and our world, even when we step away from our computers and put our phones in our pockets. Digital technology habituates its users to see things and treat things in digital ways. If they have a particular way about them—if the digital is not just a label capriciously applied to certain machines but a specific kind of technological engagement with the world—then surely the digital's way is now also one of humanity's foremost ways of relating to that world.

Here's an easy example. The idea of communication means something different in the digital age. This is trivially true: when one of my students thinks about getting in touch with a friend, she probably thinks about a text message, whereas thirty years ago, one might have thought about a phone call or a chat over coffee. Of course it's not just the thinking that has changed, but the doing. We write differently from how we used to. We gather news, form opinions, plan our lives, and purchase stuff differently. We entertain ourselves, view images, listen to and make music differently. We build things, store things, send and receive things differently. And the digital is the common factor that has made all of that difference and plenty more.

None of that would be a cause for concern, however, if it were only a matter of the adoption of a new device to replace some old ones, if the computer and smartphone were just *upgrades* of telephones, televisions, cameras, postcards, boardrooms, shopping centers, playing cards, calendars, billboards, saxophones, and so on. But that's where this book's argument gets its bite. For, beyond just extra convenience and a small form factor, digital technology has a pervasive way of being, a *digital* way of being, that structures thought and action in its image, shaping not just the tools we choose for communicating, planning, paying for things, and the like, but even the content of that communication: what gets written, how we understand what we read, what it means to be a friend, what we choose to purchase, how we imagine our futures. That altered way of thinking and acting, learned from the digital environment that now surrounds us, is so much a part of our lives that it even steers our relations to the world outside of our digital machines.

Well, where is the harm in that? Thinking and acting digitally seem like reasonable and effective ways of going about one's business. Habituated to digital devices, we may tend by default to break tasks down into discrete steps, because that's how our devices teach us to understand action. We may be less likely to experience the world as filled with uninterrupted continua and more likely to see it as made of individual parts, for that's how computers construct the objects they present to us. We may be more apt to treat the people around us as pieces to be arranged to achieve some end, since digital devices offer us exclusively means and the ends that those means can achieve. If we increasingly meet the world under the influence of digital machines, and if those machines share a common way of being, then we are becoming overwhelmingly practiced in that digital way of being, learning to encounter the whole world, on or off a digital device, in the image of the digital.

This book offers two arguments that work in tandem. One argument examines the digital as a principle of technological operation, exposing the assumptions built in to *every* instance of digital technology. For instance, a binary or two-valued code is the (practically) universal language of all digital machines, but those machines are thereby constrained to operate only on that which can be adequately represented using a binary code. The scope of this technophilosophical argument is therefore as broad as the spread of digital technology itself, for it rests on the digital-ness of digital technology, that which is common to every technology that operates on the basis of the binary code. The other line of argument focuses on the ideology or cultural conditions, the prevalent values that have encouraged digital technologies to fit into human life so readily, to become so popular so rapidly. This loosely historical argument further demonstrates that those prevailing values that promote digital technology are amplified by the very technologies they promote. This feedback circuit generates still greater adoption and application of the digital, tending toward a digital hegemony, wherein the digital (and its associated perspective) become the dominant frame through which we understand the world and the people in it. These two arguments, one technophilosophical and one ideological, meet and support each other: the examination of principles of digital operation shows that the values allied with the digital are inherent to and inalienable from its technologies, while the fertile ideology that recommends digital technology explains how the digital could enjoy unprecedented success in spite of its limitations as revealed by the philosophical analysis.

Readers familiar with current research in digital studies may already recognize that this book's approach is unusual. Many scholars of the digital investigate the diverse impacts of some particular software, or the repercussions of a new hardware platform, or the community that develops around a disruptive innovation in digital technology, but the most general question—the effects of digital technology per se—is too diffuse to attract much scholarly attention. The accepted method in digital studies of examining particular technologies in relation to their social contexts typically proceeds as cultural theory or situates itself in the closely related field of media studies, taking actual events, demonstrable historical dynamics, observable human behavior, and real artifacts as the primary objects of study and the chief evidence for any conclusions. How does embodied difference show up in our digital lives? How are race, gender, ability status, geography, class, generational identity, and so on differentiated in and through digital technology? How are power, money, labor, truth, and desire distributed and recast in relation to digital media? Such inquiries most frequently concentrate on particular technologies (smartphones, word processors, spam, the integrated circuit) or specific historical-cultural configurations (the rise of Web 2.0, GamerGate, differential access to the internet in the developing world). By contrast, this book derives its understanding of digital technology first of all from the technological principles that underpin the operation of those technologies and mark them as *digital*. As a foundational study of digital technology, this book yields significant implications but not overt recommendations for computer (and software) design and use, but it is foremost a work of philosophy rather than cultural theory, as it focuses on the nature of the digital in itself and not as it appears in some particular time or place. My conclusions thus apply to *all digital technology* and to *everyone who lives to some degree under its influence,* which is just about everybody in the world. (When this book uses the first-person-plural *we* or *us,* it means precisely everyone, to the degree that they live in the shadow of digital culture.) This is not a book that recommends a strategy for designing better human–computer interfaces or offers a criticism of the conceit of big data, though a foundational grasp of the digital will surely inform any thinking along those lines.[1]

The Digital and Its Discontents borrows its title from Sigmund Freud's influential monograph *Civilization and Its Discontents,* which applies a psychoanalytic framework to society at large, postulating that, in order to live together peacefully, human beings must necessarily leave some of their desires unfulfilled, leading inevitably to a measure of unhappiness in every society. Though the parallels between the two books are tenuous, my book reproduces by approximation the structure of Freud's discussion: an abstract, ahistorical analysis (civilization invariably causes a degree of human misery / digital thought is missing something vital to human being) that buttresses a second, historically specific argument that explains why the (ahistorical) problem takes on a particular urgency at this moment. Freud diagnosed the ills of post–World War I European society, whereas this book attempts to document the real loss that accompanies the massive propagation of digital technologies over the last half century.

Though his psychoanalytic schema was novel, Freud was not the first to recognize that living with others requires some individual sacrifice. Likewise, this book, though its approach is unusual and inventive, is not alone in launching a criticism of digitality writ large. There exists a small collection of writing from the last fifty years or so that considers the digital in general as an object of philosophical inquiry. Hubert Dreyfus's landmark 1972 book *What Computers Can't Do: The Limits of Artificial Intelligence* introduced me to this theme, and his Heideggerian argument about the limits of computation remains potent and highly relevant fifty years later, notwithstanding dramatic advances in digital technology in the interim. No less incisive than Dreyfus's exploration of the constraints of artificial intelligence, M. Beatrice Fazi's 2018 monograph *Contingent Computation: Abstraction, Experience, and Indeterminacy in Computational Aesthetics* contests the digital's reputation as uncreative and predictable, arguing that there is an essential unknowable aspect of digital execution that opens onto novelty and defies prediction. Fazi's brilliant analysis anticipates and rebuts my argument before I had even written it, and my book's claims are sharper and its vocabulary more precise thanks to her forceful vision.

Fazi's is a book of academic philosophy, and Dreyfus, though he targets a broader audience, also engages in unabashed philosophical discourse. Avowedly academic in tone, *The Digital and Its Discontents* is not a crossover book written for a lay audience, but neither is it pitched at specialists. It assumes a reader's casual familiarity with computers and smartphones and what it's like to use them, but relies on no prior knowledge of digital studies, nor philosophical training, nor technical expertise, for it offers plain explanations of both its technological and its philosophical investments. Standing apart from the digital studies canon and attempting to explain itself without presupposition, this book also includes only sparse citations and occasional references to previous work. The language is often abstract and consistently dense, but I hope always clear and direct. It should make sense to any careful reader, but will likely be of interest primarily to those for whom the question of the digital already feels worth puzzling over.

In keeping with its philosophical approach and departing again from familiar conventions of digital studies, this book's core distinction is not between *digital* and *analog* as two kinds of media or two modes of representation. Rather, the digital in this book is juxtaposed to (what I call) the *actual*; the digital and the actual are two aspects of reality with different ways of being, different *ontologies*. In many discussions of the digital, *actual* meets its antonym in *virtual*, but that latter term connotes an ersatz or lesser reality, and so forecloses a question that this book insistently keeps open: what sort of reality does the digital offer us? In contrasting *digital* and *actual*, this book acknowledges that both are fully real but emphasizes that they are differently real, and the nature of that difference stands as this book's central concern. Importantly, the essential critique is *not* that the digital falls short of the actual or fails perfectly to simulate it. Even when we call on digital tools to provide a sensorium or some other access to a simulated environment, there is rarely an expectation or need for that simulation to be indistinguishable from ordinary worldly experience. In fact, the digital works so well in part *because* it differs from the actual, with fewer moving parts, as it were. It may be a long-standing dream of media that they will overcome their mediality to offer an equivalent of the real. But there is little need for another analysis showing that digital media are not equal to the reality that they (often) represent. Instead, this book suggests that the problem with the digital is not so much an inadequacy in comparison with the actual, but an efficacy so profound and so appealing that it has become a significant component of our human reality. That nearly ubiquitous engagement with the digital leads us, in turn, to think of the world, even apart from digital media, according to values and expectations derived from the digital. And that is the problem, because there is indeed, I argue, a crucial shortcoming of the digital relative to the actual, not about the inadequacy of simulation, but about the digital's perfection: unlike the actual, the digital excludes accident, disallows spontaneity, and so minimizes that crucial dimension of actuality that this book calls "contingency." Beguiled by the remarkable capacities of the digital, we are increasingly devaluing contingency and embracing digital methods, a trade-off so satisfying that we rarely notice it at all and remain blithely unaware of contingency's perilous absence.

One risk inherent to the study of the digital in general is the specter of technological determinism. Technological determinism at its most severe proposes that certain technologies have an invariant effect in any culture into which they are introduced. Such theories are widely disparaged as untenable (and even intellectu-ally irresponsible), for they fail to take cultural difference into ac-count and fail also to acknowledge the inextricability of technology and culture. That is, most historians, cultural theorists, and media scholars will insist that no understanding of societal change, includ-ing technological change, can proceed without specific attention to the distinctive nature of each society and the varied ways in which they integrate technologies. Automobiles mean something differ-ent in Los Angeles from what they do in New Delhi. Psychoanalysis enjoys a different reception in London from Beijing.

One reason that humanistic digital studies has, arguably, coal-esced around a cultural-theoretic model and avoided the grand question that prompts this book—what have digital technologies wrought?—is a learned allergy to technological determinism. To study a particular technology as historically situated, for example, connecting it to developments in its surrounding culture, is to let the empirical lead the theoretical, eschewing a speculative char-acterization of technology and its effects. But to derive the con-sequences of technology from its universal forms rather than its practical application is to suggest that technology, in this case digi-tal technology, has an invariant effect and is thus a first cause of social consequences. And surely this is a poor assumption, because technology and culture are not strictly separable and each is shaped by the other, so that no first cause should be identifiable.

Inasmuch as it assigns universal values to (all) digital technolo-gies, this book indeed flirts with technological determinism, invit-ing the just criticism of historians and cultural theorists. But the argument herein does not finally treat culture as simply reacting to the inevitable influence of technology. Rather, digital technol-ogy in this analysis gains its cultural authority because its associ-ated values were already prevalent in Western cultures when digital technologies began to materialize. A culture not already under the sway of the ideology associated with the digital would integrate those technologies very differently, if at all. Further, even in cul-tures that are ideologically ripe for the ascent of digital technology, its positivist intensity tends to provoke not universal conformity, but partial reactivity in which some people recognize and attempt to restore the ways of being that are excluded from the digital, a revanchement of the vinyl record, for example, or a recommitment to face-to-face human interaction. Thus, the purported determinism of the method of technological analysis in this book is mitigated by the generality of its conclusions. Digital technology, I argue, points everyone in a certain direction, but there are many paths along that heading, as well as plenty of travelers who drift off of those paths to move in other directions altogether.

Algorithms are a popular point of focus for much digital studies, because they play an agential role in the digital, and can thus plausibly be assigned responsibility, if not ultimate responsibility, for what the digital accomplishes in the world. If a corporation exercises its power via a digital medium, it is usually sensible to understand the algorithm as one prominent means or instrument by which that corporation wields its power. Algorithms are a vitally important organizational structure within the digital, and digital scholars are wise to treat them as primary objects of study. But to focus on the algorithm is to give other points of focus less attention. For instance, plenty of scholars study digital *data,* what it is and what one can do with it, and this too is an essential query in the digital age. If this book takes a still broader perspective, asking about the digital in general, this is not to ignore the importance of algorithms or data, but rather to allow the philosophical method herein to lead where it will. Some might regard the breadth of that philosophical inquiry as disqualifying, claiming that *the digital* is simply so nebulous and so complexly entangled with culture and history that it evades a specific examination. As mentioned, this book attempts to surmount that hurdle by apprehending the digital through its underlying technological principles, the invariant features of digital technologies that make them digital. In particular, it recognizes *the bit* as the universal element of digital technology and then asks what can be done with bits and what possibilities are discouraged, or even unreachable, via bits. Focused on this general analysis, this book often appeals to hypothetical rather than actual examples to illustrate and interrogate the ideas it discovers, making only occasional references to specific platforms, corporations, protocols, and digital cultural practices.

Following this introduction, the book's chapters lay out the argument slowly and deliberately. Chapter 1 extends the work of this introduction, offering an overview, identifying appropriate readers for the book, introducing some key terms, and discussing the stakes and style of argument. Chapter 2 frames the central questions that this book addresses, articulating the book's motives and then providing an account of the ideological commitments associated with the digital. Chapter 3, the most philosophically intense, sets out the ontology of "the actual" to heighten the subsequent contrast with the ontology of the digital, calling on some capsule versions of relevant philosophical ideas and detailing the actual's ontology through an extended figure of a mesh out of which reality is constructed. Chapter 4 is the most technical chapter, presenting the ontology of the digital by examining the bit, what it's made of, how it works, and what one can do with it. Chapter 5 builds on the previous chapter's analysis by looking at how bits are employed in digital machines to connect those machines to a human world and make them useful. Both chapters 4 and 5 emphasize the digital's differences from the actual, identifying the former's mechanisms for warding off contingency, which make it both so effective and so problematic. Chapter 6 considers the implications of the digital ontology for the people who live in the midst of digital technologies and bear the stamp of the digital in their actions and thoughts. Though this book includes counterargument at many points throughout the text, the final chapter, chapter 7, collects some of those counterarguments and attempts to look directly at the apparent contradiction between this book's account of the digital as uncreative and static and the heady excitement and sense of novelty that has accompanied computers and other digital devices for almost a century. As a concluding gesture, this last chapter also looks toward the possibility of taking advantage of the remarkable conveniences and capacities of the digital without succumbing to the threat of the reductive and moribund ideology so closely associated with it.

‹ **1** ›

Approaching the Digital

Like so many others, I love digital technology. I was an "early adopter" before that was an established term, learning to program while I was in middle school in the late 1970s, and purchasing a used Apple II Plus computer around the same time, at the outset of the personal computer age. It's not just that that computer made it easier for me to write and edit high-school essays or play games late at night when the rest of the household was asleep; my attraction to the machine ran deeper and was more emotional than its utility would warrant. Basically, I thought the computer was impossibly cool, the shiniest kind of high technology, and one could see already by 1980, walking through the mall or exploring the science wing of a high school or just reading science fiction, that computers were going to be an increasingly huge part of the future.

That intense libidinal investment in digital technologies has not faded for me more than forty years on and is now shared widely across the globe. "Unboxing" videos on YouTube, long lines waiting overnight for the latest cell phone, a race for ever tinier circuitry on silicon chips, the massive growth of the video-gaming industry, the widespread reverence for TED talks—these are symptoms of the boundless desire for the digital, and the mad enthusiasm that surrounds digital artifacts belies their commonness as anodyne hunks of metal, glass, and plastic. Whence this seemingly excessive attraction to digital technology? What's so great about computers?

That's the question that drives this book. How and why has the digital become the default choice, a pervasive addiction, the answer to all of our problems (and the source of many of them), an inexhaustible font of futurity? Digital dependency is not only personal, not just about a need to check Instagram or Facebook or email. Corporations, governments, and institutions of all sorts choose almost every time to embrace digital technologies, often acting on a faith that it will be somehow better, that the digital is the answer. The question of the digital's appeal becomes still more perplexing, even urgent, with the recognition that, alongside our headlong desire lies an uneasy suspicion about computing technologies. Even as we turn more and more often to the computer and smartphone to conduct our daily lives, we can feel something superficial there, a simulacral quality, but still, notwithstanding the popular wisdom of "unplugging" (for a little while) and plenty of handwringing about lost handicrafts and lost languages, we typically ignore any reservations and gleefully click the button, assured that the package will get expedited shipping. After all, one wouldn't want to fall behind.

Engaging questions about value and about culture, this is therefore at once a book of philosophy, an analysis of technology, and an examination of ideology. It advances two parallel related arguments about the digital. One line of thought exposes a powerful *digital* ideology: not just the efficacy of digital technology, but a set of widely shared values, growing in influence since the Enlightenment, has supported the rapid, blanket spread of digital technologies around the world. Buoyed by those prevailing values, the digital enjoys an uncritical acceptance and an often giddy enthusiasm that began even before the digital's rise as a corporate, industrial, military, and personal technology. Actualizing those values in its most basic principles of operation, the digital also promulgates them, feeding a closed circuit that accelerates the adoption of those values and encourages concordant habits of thought and behavior, which intensify, in turn, the desire for the digital. When other, *nondigital* ways of thinking are squeezed out to become scarce or unavailable, digital values come to seem natural and inevitable, as though they represent the only possible relationship to the world, a textbook example of ideology at work.

Though this book leans rather toward philosophy than history, its critique of ideology refers to the way that the digital changes, and changes us, over time. The other line of argument examines the digital as a general feature of *digital* technology, identifying the core structures of those technologies, structures that do not vary in their essence and are thus shared by all digital technologies in every context. In other words, this argument proceeds by analyzing and evaluating that which makes digital technologies *digital*. Showing that the universal principles of the digital inevitably demonstrate and promote the values identified by the analysis of ideology, this book claims that the relationship between the digital and those Enlightenment values is not an accident of history, but rather an inevitable artifact of digital operation. It is only because of its inalienable alliance with those values that the digital can be such an astoundingly effective tool and can play an essential role in so many human activities today. But those same factors of digital operation that make its associated technologies so powerful and popular also carve out a fundamental deficit in the digital, an impoverished relationship to *contingency,* with subtle but widespread and worrisome consequences. Because we share the computer's values, and increasingly practice its ways of seeing and acting even when away from the machine, we tend not to notice what's missing from the digital, and redressing this oversight is the chief charge of this monograph.

This technophobia may sound unduly paranoid, but the claim that there is something problematic or destructive about the digital is hardly original. Digital technology has occasioned a great deal of suspicion and consternation, though those worries tend to be more intuitive than reasoned. Digital machines are accused of being cold or sterile; they dehumanize or distract; they pervert sociability and mechanize thought; they promise a perfect world but deliver only shallow simulacra. In any case, one often hears, computers are not to be trusted. Popular fantasies like *The Matrix* or *Black Mirror* exaggerate this digital distrust, depicting a revolution in which the machines enslave their makers, or erect a surveillance state from which one cannot hide, or institute a prejudicial system of algorithmic selection unchecked by human ethical consideration. This book identifies the source of the digital's strengths and its weaknesses, explaining why the digital provokes such uneasiness and showing that the real threat is not so much a robot uprising but, more subtly, the possibility of losing touch with important parts of ourselves and our world, as digital modes of relation eclipse other comportments. Digital technologies satisfy a desire that preexisted them, but also inculcate and inflame that desire, a feedback loop that populates the world, including the people in it, with the digital and its exclusive values. (So, yes, a kind of robot uprising.)

In other words, the digital gives us what we think we want. It adheres to an exemplary *positivism,* a devout *rationalism,* and a compliant *instrumentalism,* which are entwined in a pervasive ideology that binds the digital to capitalism, liberalism, and other hegemonies of the present day. These values, herein labeled *digital* values, are part of the background account of the social milieu and invite many angles of analysis, so that one contribution of this book is to frame these post-Enlightenment values in their specific relation to the digital. Positivism, rationalism, and instrumentalism are not born from digital machines, but their increasing prevalence in recent centuries explains how digital machines could come to seem, in the middle of the twentieth century, like the solution to so many challenges and the best way to get things done. It is tempting to diagnose the threat of the digital as a dehumanization: one might say that the digital lacks something essentially human, such that we risk losing our humanity when we submit to its procedures. But the values associated with the digital (positivism, rationalism, and instrumentalism) are originally and characteristically human, and in fact they accord rather well with the legacy of humanism that was also accelerated in and after the Enlightenment. To calculate, to reason, to employ means in the service of ends—these are essentially human ways of behaving, though they may not be uniquely human. The computer's foremost deficit is therefore not so much an insufficient humanity, but something still more basic, a poverty of *contingency* that divests the digital of a richness and plenitude shared outside the digital by humans and by the nonhuman, across all scales of time and space. It is the digital's rejection of contingency that entwines it with positivism, rationalism, and instrumentalism, and thereby also spreads those values of thought and action among those who fall under the digital's influence.

To be clear, this digital ideology is not first of all a political ideology, and its analysis in this book is not primarily social critique. As a mode of relation to the world, the digital ideology undoubtedly has significant social and political repercussions. Many fine scholarly, journalistic, and speculative works examine the sociopolitical orders that accompany and appear to derive in part from digital technologies and media. (Examples abound, but the canon would surely include research by Wendy Chun, Manuel Castells, N. Katherine Hayles, Lisa Nakamura, and many others.) This book does not reproduce or parallel those analyses, but aims to support them by showing the tight link between the digital and the ideological schema that drives so much thought and action today, generating ways of being that constrain available possibilities of social organization. For example, conceiving of the value of persons in positivist terms reinforces social hierarchy by sustaining the illusion of a natural order. But whereas most critique of ideology bears on the way that it preserves or institutes social structures or political hierarchies, the present examination of digital ideology focuses on its values (positivism, rationalism, and instrumentalism) as practiced by individuals, rather than spelling out the consequences of those values for the organization of groups of persons.

To demonstrate the link between digital values and a diminished contingency requires an examination of digital technology in some technical detail. The digital's exclusion of contingency and, so also, its ideological complicity are manifest in its core principles of operation, the technical details of *digital* machines, accounting for its remarkable affordances at the cost, rarely noted, of a crucial deficit. Those underlying principles effectively banish contingency by imposing rules everywhere throughout the digital, leaving no room for the free play, creativity, openness, or accident that are contingency's content and expression. And so this analysis employs at some points the language of engineers and programmers, but these technical details, like the digital in general, are not finally very complicated. Digital operation is at bottom pretty simple, though it is a simplicity often belied by the nearly inconceivable number and supersensory speed of digital processes. By contrast, the complementary analysis, the examination of the vital role of contingency in the actual, is more contentious, more subtle, and more difficult to demonstrate. Chapter 3 employs an extended metaphor, the world as "mesh," to articulate the essential role of contingency in both human and nonhuman aspects of our world.

We traffic so often in positivism that we rarely think of what might resist a positive account, and so we fail to notice many of the ways in which the actual differs from the digital. This book will be compelling only if it effectively critiques an ideology so widespread as to have become invisible. By the force of a habit of thought already aligned with digital values and the implicit belief that there are no other values, the digital's flaws are not readily demonstrable, for we are primed to ignore them. Taken individually, an example or even a direct argument might seem anomalous or inconsequential when viewed through this ideological filter. Given that the digital is so effective, does it really matter (a reader might ask) that we use predetermined algorithms to animate a face or rely on statistical analysis to measure a health risk or construct simplified digital models to predict the path of a storm? Such digital practices are so useful, so apt, that their value seems immediately to trump a subtle and abstract worry like a diminished contingency, setting the argument of this book on a steep uphill climb. This monograph therefore attempts to pry open another path for thought via a supplemental mode of argument, interrupting the continuous text with a series of "vignettes." These vignettes motivate questions by introducing examples and analyses that expand but also challenge the book's core argument; no single example is decisive on its own, but these interludes accumulate a force that aims in aggregate to pierce the stubborn veil of ideology.

Digital gaming offers objects of inquiry for a number of those vignettes and pops up elsewhere as part of the central argument throughout the text. This is not, however, a book principally about digital games, which are employed herein as privileged examples to illustrate and interrogate digital phenomena more broadly. Rather, gaming is prominent in this analysis because it provides a perspicuous view of how the digital works. Games usually require an uncompromising focus on the interface between the player and the machine. Whereas many other kinds of software serve as means to ends that are ultimately outside of the computer, digital games are often purely digital media, an activity of and in the digital machine. Given that many games also place a maximal demand on the machine's resources and tax its most important subsystems, gaming tends to expose most explicitly the underlying logics of the machine and the mechanisms by which designers and programmers attempt to overcome its limitations. They therefore offer an excellent laboratory in which to study the digital generally, as they both exhibit its ideology and, with great success, disguise that rigid ideology behind a conception of "play" and its closely associated sense of freedom.

The argument does not discover this powerful digital ideology as a habit of thought arising in the age of the digital machine, but rather tracks an intensification, via the digital, of long-standing relations to the world that are increasingly prevalent since the Enlightenment. That is, it is a question of degree: ideology's rule is not absolute, but shapes a greater or lesser share of our world. Ideology is belief that eludes explicit awareness, the set of shared, unspoken assumptions that more or less determines what counts as proper, or valid, or valuable. Its effects are subtle and nebulous, even as digital ideology weighs against subtlety and vagueness. As such, there is plenty of evidence showing that we have come to understand the world through a digital lens, but also plenty of evidence demonstrating our persistent valorization of the antipositive continuity that the digital cannot capture. For every advertising image of a car engine or fruit orchard or medical record breaking down into a positivism of (digitally) individuated parts or pixels—ads that work on the assumption that mastery over data and mastery over the things that data represent are largely the same thing—there is another commercial showing a continuous flow from the earth to the refinery to the consumer's home, or an animated *mash-up* of one species of animal transforming smoothly into another, or a panoply of athletes suggesting a stirring and uncategorizable diversity of humanity, or an uninterrupted transition from ancient civilization to the far future, images of field and flux, disindividuated collections of heterogeneous or manifold substance that do not resolve into independent bits. Even as we increasingly embrace the discreteness at the heart of digital operation, we also yearn for the connection and immersion that characterize our extradigital world. (In the original film *The Terminator,* the eponymous machine is a fleshy skin stretched over a highly articulated collection of individual metal parts, whereas subsequent terminators leverage the antipositivist threat of infinitely malleable plasticity, a flow of liquid metal without fixed identity.) Positivism has its means of taming continuity or flow, say by assigning a measure or taking regular samples, but positivism, a relation to the real conceived as discrete and autonomous *posits,* can only simulate flux, but never finally capture it.

Rationalism too has its spheres of influence, establishing a discursive default, such that, in contexts that call for an explanation, the very act of explanation depends on a broadly shared notion of what's rational. Behaving according to reason, or just being reasonable, is a minimal and universal norm. But rationalism, like positivism, is respected as much in its breach as in its honor. Who would insist that one's preferences regarding music or food or favorite colors should be rationally defensible? Aristotle maintains that human beings are distinguished from other animals by rationality, but he devotes much of his studies of ethics and politics to the ways in which people fail to exercise reason in their real lives. It is even a source of pride to maintain an idiosyncratic or indefensible position on some matters, as though it were a sign of the singular strength of one's character, the faithful belief that is its own reason, invoking no other.

And if instrumentalism likewise serves often as a presumptive baseline, such that utility or means–ends reasoning is a default mode of evaluating the world (objects and events), then it is also true that everyone, in body and soul, expends much effort in the service of no end: collecting things of little value, beginning projects they won't complete, twitching and muttering, tossing and turning to achieve nothing but the perpetuation of a *useless* habit of being oneself. Like positivism and rationalism, instrumentalism's ideological force bears most heavily on what we say, on how we talk and think about things, but the world in general does not obey the mandate of instrumentalism and so derives its values not only from utility, but from all quarters. Even thought, or really especially thought, though its articulation may be subject to the guiding norm of digital ideology, wanders widely when loosed from the grip of language, ideology's eminent domain.

It is in language that we reveal to ourselves most explicitly what and how we think, but the exposure of thought in expression gives ideology its foremost opportunity to enforce its social norms. As ideology's compulsion peaks therefore in language, as it governs first of all what one is supposed to say, its defiance can lead to some strange talk. Among the unlikely premises advanced by this book are commitments to the indeterminacy of things (including people), the nonequivalence of a thing with itself, the priority of becoming over being, and the rejection of a bottom line, of what some might call "Truth with a capital 'T.'" Relatedly, this book's argument asks readers to embrace a spontaneous reason, a sense-making in which the rules are not determined in advance, but are given as the sense being made; reason might make any appeal, might in the right context overthrow any particular rule, but this is not to say that just anything at all is reasonable. Reason makes its own rules, but not without its reasons.

Whether these petitions meet with accord or dismissal depends of course on the cogency of the arguments herein, but also on the reader's metaphysical bent. Though it is not the only available creed, we are taught to believe, at least when pressed, in a world of real, discrete, stable things, in a self (or soul) that stays the same amidst changes "on the outside." These posits—things, selves, atoms, propositions, facts, and many other positivisms—establish a basis for truth and falsity, rules for what is reasonable or senseless. Those who are satisfied with this account of things, people who prefer the solid footing of that serifed capital "T," will be disinclined to credit the critique of this work and less likely to share its worry about a fundamental problem in the digital. By contrast, those who live in a world of difference, welcome the sudden and spontaneous, admit no first or final truth, delight in accident and indeterminacy, those who embrace *contingency*, will find here a resonant articulation of a familiar doubt regarding computers and their placement around and among us.

Engaging a digital metaphysics or *ontology*, this book makes an unusual, though by no means unique, intervention into critical discourse on the digital. Humanist scholarship about the digital usually borrows its approach from cultural-theoretic disciplines, treating the digital in relation to its historical and cultural contexts.[1] This book joins recent research in *digital philosophy*, a field that examines the general nature of the digital at a degree of remove from particular historical and cultural situations. Specifically, *The Digital and Its Discontents* discovers in *the binary code*, and the way it is put to use in digital machines, the mechanism of both the digital's astonishing capacity and its subtle incapacity. Bits, the universal elements of digital technology, have two sides: they represent the abstract code of 0s and 1s but also concretize and materialize that code so that it can act in the world. They work by being simultaneously symbol and substance. It is this conjunction of the abstract and the concrete, a purely formal code rendered materially effective, that empowers and delimits the digital. In a nutshell, *the digital works by virtue of its formalism.* The unfolding of that claim distinguishes most sharply the formalist analysis of this book from the bulk of research in digital studies. Moreover, a philosophical account of the digital grounds an understanding thereafter better prepared to study the digital's empirical and social dimensions.

So, what to recommend? What shall we do? The digital amplifies and spreads the associated ideology, but it is that ideology that props up the digital. A contingency rendered mute or meager might therefore be revitalized by turning away from the digital, but effective resistance must aim first of all at the underlying digital *values.* Hence the critique of ideology. This book will have succeeded if the reader can more readily recognize the digital's shortcomings lurking behind its dazzling feats of calculation and rendering, its undeniable convenience, its frenzied pace and inconceivable number, its apparent ubiquity, laying claim even to our futures. To bear witness to what can never be digital, to prize the contingent that destabilizes posits, imposes a new reason, and frustrates utility— that is this book's recommendation and its hope.

Imagine a Computer

Think about a computer without the usual material limitations, a computer that calculates arbitrarily quickly, that has as much available memory as it ever needs. This computer can be programmed, say, by a focused act of intention: think it clearly enough, and the computer will do it for you. Its output bypasses the fixed limits of screens and speakers to effect a direct address of the nervous system, overcoming issues of resolution, field of view, and flatness and including the full human sensorium. Unburdened by material resistance and demanding the minimum of intellectual labor, this computer would still operate, we are imagining, on the basis of a binary code, still rely on a sequence of unambiguous commands to manipulate various data to arrive at some result. Absent constraints of time and resources but still bound to the binary, what limitations, if any, would such a machine encounter? If we are no longer worried about resolution, no longer obliged by the work of software engineering, freed from material concerns, what could such a machine accomplish and where might the power of the binary code still fail to reach?

What Does the Digital Do?

Diagnoses of our contemporary condition frequently open by observing the manifold changes in human life since the rise of digital machines. The digital has altered our world and our ways of being in it: in the fine details of our quotidian affairs such as how we sleep and eat, work and shop, relax and socialize, but also in matters of grand or global significance, such as politics, finance, science, and war. What domain of human activity has gone untouched by digital technologies? Even writing, even money, even electricity have an applicability more defined and delimited, less disparate and sprawling than the digital. The digital recasts our broadest horizons, mediates much of our everyday commerce, and abides in the intimate and subtle corners of our lives. Enmeshed in the gamut of human activities, the digital appears to erect another world alongside of *and within* our own. What sort of world is it?

Scholars, gurus, industrialists, artists, engineers, psychologists—they pursue different paths toward an understanding of the digital. But for all they reveal from one or another perspective about Microsoft, memes, design patterns, automated labor, word frequencies, interfaces, or the ethics of algorithms, seldom do these voluminous discourses address the digital per se, asking after its general nature, its way of being. We could list thousands of applications, things the digital can be used for, but that would only beg the question of how this singular technology does all that different stuff. What does the digital do? And how does it do it?

Many likely reject the premise that there is *a* digital and that it has *a* nature; for them, the digital is a syncresis, a label to apply

ad lib to lots of the phenomena that characterize our highly technological moment. But nominalism in this case offers no explanatory
value, and moreover runs afoul of the evidence. For, the digital is
well defined and even surprisingly simple: what is digital is what
works by way of discreteness. In the case of digital technology, its
essential feature is the discrete binary code, a code of 0s and 1s, by
means of which the digital stores *all* data and performs *all* operations. The rest of this book makes the case that this two-valued
code, the universal and definitional principle of digital technology,
is highly consequential, responsible both for the unprecedented capacities of the digital and for its mostly unremarked flaws. What the
digital does it does by virtue of the binary code.

It's not only that the general idea of a binary code, lists of 0s and
1s, is uncomplicated. Even the technical details of how this code
makes possible all the manifold things the digital can do, even that
more thorough explanation is fairly straightforward. For a technology that appears opaque to so many who use it, its principles of
operation are relatively easy to understand. A computer is a powerful calculator: all of its data are numbers (0s and 1s), and all of its
actions are calculations with those numbers, arithmetical and logical calculations. The digital processes numbers using addition, sub-
traction, multiplication, comparison, and other fifth-grade-math
superpowers. (The computer can calculate somewhat faster than
a fifth grader.) Because all the data are turned into numbers, and
because all the numbers are turned into sequences of 0s and 1s, the
elemental operations of digital technology are never more complicated than two-valued arithmetic problems: 1+1, say. The primary
prerequisite for an operation on a computer is that the operation
and the data on which it operates be first rendered as sequences of
0s and 1s.

Simplicity is not transparency; a proprietary algorithm may include factors known only to its designers, and even accessible information about what goes on behind the scenes of a web page or
video game is meaningful only to those with the specialized knowledge required to interpret it. But the general principles that describe how digital machines work are eminently comprehensible
to the nonexpert. Without knowing the particular algorithmic
methods that govern a given search engine, one still knows or can

easily come to know that, as a digital algorithm, it performs statistical calculations on a set of numerical data that encode occurrences of words (or images, etc.) in the data to be searched. That statistical analysis yields a number of (numerical) maxima, and references to the data associated with those maximal values are output as a ranked list. The search results thus depend on how those maxima are calculated, what factors are included and how they are weighted, which might involve some heavy mathematics as well as some ethical responsibility. But the contribution of the digital algorithm is just to carry out those calculations faithfully, reducing them all to the simplicity of binary arithmetic.

To define the digital as what is grounded in discreteness extends the term well beyond the technology category that has labeled our age. With its two discrete states of *on* and *off*, a light switch is conceptually digital, even if its lack of a code excludes it from the normative notion of a digital technology. Typically, technologies are digital, because they employ a discrete code as the mechanism of their operation, but the digital as the idea of a hard distinction (between one thing and another) is an originary dimension of human thought and action, a basic part of our world. Chapter 3 will argue that the world itself, apart from human subjects, includes distinction, but that the ideality of absolute discreteness is an artifact of human thought, which draws to an extreme the world's suggestion of approximate and partial differences among things. In any case, digital technologies elevate discreteness to a central operating principle, but they do not invent it. The digital, in this broad sense, can be found everywhere. Counting, for instance, counts as digital, even when it is not notated using digits, for it identifies and distinguishes one individual thing from another. (Fingers and toes are archetypical digits.) Alphabets and pictograms, language, money, property, gender, the distinction between true and false, what is and what is not—discreteness seems foundational to thought itself, coeval with the human. Or, rather, the world points toward discreteness from the start, and humans (but not only humans) encounter it as a constant of experience.

For the most part, "the digital" as used in this book refers specifically to those technologies that bear its label. But, as noted here, there is also a primordial relationship, a *bearing* toward the world

that could anachronistically be called "digital," one that long pre-
dates the computer age and is arguably a universal facet of human
experience. Chapter 3 hints at the connection between this atavistic
digital thought and its intensification into the more recent ideological forms critically examined in this book. That connection helps to
focus the critique, which targets not the practice of thinking about
things as discrete but the ideological purity that grows out of this
practice. The polysemy of *the digital* should thus be heard herein,
but only faintly.

"What does the digital do?" has been understood at the start
of this chapter as a question about how digital technologies work,
but we might also take it to be asking about the digital's effects, its
accomplishments, its capacities. Indeed, to discover that the digital's mechanism is so simple only confounds the question of what
it does, for it is less than obvious how the use of a binary code in a
very fast calculator could be the basis of the most widespread technology in history. A list of things the digital does—spreadsheets,
threat detection, archiving, computer-aided design (CAD) and
computer-aided manufacturing (CAM), online poker, photogra-
phy editing, accounts payable, toaster control, chat, and on and
on—evinces only the breadth of digital application, its fantastic utility. One could extract some generalizations from this long
list, attempting to summarize the digital's capacities in the hopes
of gaining greater perspective regarding what it does: calculation,
automation, algorithms, audio, video, communication and data
transfer, simulation, logic, interactivity, and so on. This makes for a
shorter list and moves toward a more satisfying explanation of what
the digital does, but these digital doings remain too heterogeneous,
too much of a hodgepodge to provide a satisfying, coherent sense
of the digital. What do these various capacities share? What is their
common ground? What does the digital do, with its sequences of 0s
and 1s, that enables it to do so many things?

Martin Heidegger's famous analysis in "The Question Concerning Technology" points toward another hearing of the question.
Guided solely by the imperative to "ask after" technology, presum-
ing at the outset no particular direction of inquiry, Heidegger examines multiple histories of technology that each lead him to the
same place: technology is one way that the world shows up for us;

it shines a light on certain human relationships to nature, including human nature. But in modernity, or roughly since the Enlightenment, that relationship has rigidified, a development that Heidegger identifies as *Ge-stell*, or in English, "Enframing."

When it comes to *modern* technology, as Heidegger does not in this essay deal with digital technology, the relationship it lights up between humans and nature is characterized by demand and domination. In modernity, says Heidegger, we challenge nature to *order* itself as ready to be used, available as raw materials or energy; nature appears to us as thus orderable, as resources to be stockpiled, which Heidegger calls "standing-reserve" (*Be-stand*). And the worst of it, says Heidegger, is that this way of seeing the world, a world ready to be broken down and ordered for availability, has become just about our only way of seeing the world. Not just nature, but everything, including people, now appears for us as standing-reserve, resources stockpiled in anticipation of their use. Heidegger concludes his essay by asking how we might peek around the edge of this *Enframing* that seems to fill our entire field of view, so that we might open up other ways of seeing.

As the current representative of technology per se, the digital fits Heidegger's critique, but not without some friction. Heidegger notes the importance of numbers in modern technology, and the digital pushes that association to an extreme, constructing its entire world, data and process, out of numbers. Moreover, Heidegger's critical emphasis on the uniformity of *Ge-stell* as the one way we tend to see things finds palpable support in the growing hegemony of the digital. Digital technologies have widely replaced others and now preside over many (and for some people, most) occasions of human sociability, creativity, inquiry, and thought and action generally. The digital mediates our relationship to the world, and so if it promotes a worldview, surely that worldview is now also ours.[1]

Though it exemplifies in many ways the idea of modern technology that Heidegger indicts, the digital does not readily affirm his analysis of standing-reserve. No doubt the laptop is always at the ready, cell towers have already woven a field of radio waves ready to carry a signal around the world, and the Photoshop filter is already installed on your machine, ready to apply just the right amount of blur to your image's background. But digital technologies process *information* rather than raw materials; what they make immediately ready or available is not nature or its resources, not material things, but representations. The digital's well-known affinity for images, evident at least since personal computing's popularity, is only one of its representational modes. Digital representations are often visual but also informational, such as a representation of a thing as numerical values that measure and qualify it, a representation of a thing using words, or a record in a database, or a flow chart, or an encoded algorithm. These ways of representing show how the digital, like *Ge-stell* more broadly, orders the world to make it available, but it is an order of representations, not of the things themselves. Digital technologies' operations take place in a digital world such that they require prostheses to reach into ours. As representation machines, digital technologies are *about* the world, but only secondarily or tenuously within it.

Information is not standing-reserve, but perhaps it functions analogously when applying Heidegger's critique to digital technology. To extend this hypothesis: *what the digital does* is to establish a framework in which the world reveals itself as information. The 0s and 1s that are the ultimate native vocabulary of the digital guarantee that information is its sole object; the digital traffics entirely in information, which discovers in the digital its ideal host. Pending chapter 4's detailed examination of the binary code as a representation of information, and drawing inspiration from Heidegger's argument about technology, we now have the vital clue in the investigation of what the digital does, the start of an account of its way of being: the digital reveals the world as information.

The binary code is not innocent. Its information content can have only one form, the invariant form of sequences of 0s and 1s. This already implies a great deal about how information inhabits digital machines. Whatever is digital, whatever representations the machine makes available, those representations as sequences of 0s and 1s exhibit persistent characteristics—a way of *being*—that commit the digital to its ideological consequences—a way of *seeing*. The binary code is *discrete, mechanistic,* and *inert*; these are its characteristics that stand guard at the gates of the digital, for whatever would be digital must adopt these forms.

As has been emphasized, discreteness is the core criterion of the digital and the first characteristic of the binary code. In fact, that code is twice discrete: there is absolute distinction between 0 and 1; but also there is absolute distinction between each bit and its neighbors. Each bit has a distinct *value,* and each has a distinct *place* in sequence, and the difference between these two differences makes the code possible.[2]

To say that the binary code is mechanistic is to note that it works according to strict and unambiguous rules. Everything the digital machine does, every single process or action, must be directed by a rule that itself takes the form of binary code. The machine never performs an interpretation, never makes an intuitive leap, never even poses the question of what to do next, for its every step answers to an unassailable mechanism, a rule that determines on the basis of binary arithmetic the next state of the machine. To process information digitally is to render the information and its processing as binary calculations, a determinate sequence of 0s and 1s. Only a reason that is capturable as mechanism, executable as a sequence of binary calculations, can have a place as digital process.

Moreover, the binary code is inert. It does nothing on its own, but awaits its digital mechanism, to which it submits entirely. Mechanism means that the digital does whatever its calculations determine that it must; inertia means that it does nothing else. It has no direction but where it is inevitably led. It has no significance but what it is assigned. It has no effect but where it is heeded. This inertia, the digital's lack of autonomy or spontaneity, grounds the false belief in a digital neutrality: it does only whatever we ask of it, so any unfortunate consequence in the digital must be our own doing, it is said.[3]

Discrete, mechanistic, and inert. The binary code hardly sounds incalculably powerful or ineluctably desirable. Yet, evidently, we desire it, for we have embraced no technology with greater fervor, and we happily imagine it as a central fixture of our futures. The characteristics of the binary code point toward a set of values, an ideological nexus, that advances the digital as an exemplar of those values, and explains its hold over us, the basis of its appeal.

Corresponding to discreteness there is *positivism,* to mechanism, *rationalism,* and to inertia, *instrumentalism.* These three terms coalesce in a historical confluence that cultivates the rise of the digital. Growing in influence for centuries, these *isms* kindled a desire for digital technology, which has become their most refined expression. As a technology, the digital appears to answer to some of our dearest fantasies, offering in microcosm a version of the world precisely as we wish to have it. Positivism, rationalism, and instrumentalism together form a knot of Enlightenment thought, yielding conceptions of truth, practices of reasoning, and legitimation criteria for knowledge, action, and belief. Like all ideology, these comportments do not broadcast their foundational status, but assume the position of default, frames with no outside that become, thereby, unassailable. To challenge positivism, for instance, to assert that difference takes priority over identity, is not only an insult to the reigning liberalism of our age (where individuals are the ground level of being, whether persons, nation-states, communities, or institutions), but likely appears as a loss of reason altogether; what sense can be made when one refuses to begin with clearly existent, coherent, and unified *things*?

Positivism

The label *positivism* collects a number of schools of thought that share, for the most part, a commitment to the idea that knowledge must be grounded in facts or *posits*. Closely associated with the label *positivism,* nineteenth-century philosopher Auguste Comte most famously privileged knowledge according to the degree to which it could be grounded in the certainty of mathematics and stated as a set of absolute inviolable laws, but his totalizing positivism recapitulated the systematic understanding of knowledge advanced more than a century earlier by Christian Wolff. Those philosophers believed that knowledge could be reliable only when derived from the empirical, only when based on observable facts about the world. The twentieth-century school of thought known as *logical positivism,* central to the tradition of analytic philosophy, supported a version of positivism less invested in empiricism, emphasizing instead the possibility of rendering all truth as logical derivations from a set of minimally simple claims, or posits. Across four centuries, these theories commit to a world, or a knowledge of the world, based on very basic, autonomous propositions that are taken to be ground-level truths. In other words, they reduce the world to posits.

To posit something, an idea or a claim, is to place it before us, to establish it as given, and positivism is the insistence that all proper knowledge must be constructed from such givens. Scientists and philosophers of science have often adopted positivism as a criterion for determining the legitimacy of scientific theories: a scientific positivist believes that only those theories derived from empirical evidence can be admitted as valid, while theories that include, for example, imaginary elements for which there is neither confirming nor disconfirming evidence must be rejected as speculation and not knowledge. This understanding of positivism casts it as a principle of epistemology, a principle that governs what we can know. But more generally, positivism can also refer to an ontological commitment, particularly the belief that reality (and not just our knowledge of reality) is ultimately constructed of posits, individual states of affairs that are simply and fundamentally true. It is this ontological notion of positivism that upholds the ideology associated with the digital.

At its simplest, positivism is a belief in the ontological priority of *things*. A positivist sees the world as an aggregation of the things in it, each of which is real, individual, and in principle autonomous. Put otherwise, a *posit* is first and foremost *itself*, and only secondarily does it enter into relations. A different outlook might understand things as arising out of their relations, which would favor nonpositivist ways of being: the social over the individual, the continuous over the discrete, becoming over being, and finally difference over identity. But, even if digital thought is not the only thought today, even if nonpositivist thought still orients our worldviews in some measure, shaking identity loose from positivism's grip, still such alternative ways of seeing are treated as suspicious or fantastical, as perverse or idealistic, or are rendered otherwise benign, at least in discursive contexts. Positivism need rarely defend itself, for it is the presumptive criterion of right thought, an ontological liberalism that serves as an accepted norm.

The broad definition of the digital offered in the first chapter of this book makes it nearly synonymous with positivism. The digital is a relationship to the world that understands it as made of discrete things. Positivism also builds its world out of individuated entities, but the essential criterion of positivism is the self-assertion (or *self-positing*) of each thing, rather than the immaculate divisions among things. If the emphases are different, discreteness versus positing, the two perspectives nevertheless overlap sufficiently to form ready complements, such that a digital outlook or a digital machine, equipped to treat everything as discrete units, generally leads to a positivist worldview, while positivism nurtures the digital, welcoming its essential commitment to a bitwise construction of the world. Bits are, after all, elemental posits.

A close cousin of the bit, the atom institutes an ancient variety of positivism. An atomist holds that the world is made of small particles, each distinct unto itself and independent. Atoms do have relations to each other (molecules manifest several kinds of atomic relations), but at least for an atomist, those relations are secondary. Relations accrue around an atom, but it is first of all what it is independently of any relation. Of course, modern atomic physics, rubbing up against quantum mechanics, turns in both positivist and nonpositivist directions: an atom is made of neutrons, protons, and electrons, which to an extent function as posits themselves, but the behavior of these and other subatomic particles challenges the clear-cut, *it-is-or-it-isn't* orthodoxy of strict positivism. The ambiguity between wave and particle demonstrated by photons (and pretty much all of the tiniest bits of stuff) does not admit of a comfortably positivist interpretation. Because a positivist world is made of distinct, well-defined posits, a positivist cannot admit that there is ambiguity *in the real*, but could allow at most that some things *seem* ambiguous due to our lack of knowledge or understanding. Real ambiguity, which may be inherent to quantum mechanics, is positivist anathema.

Physics of Imprecision

The physical determinism of classical mechanics has long seemed unassailable: if the energy and matter that constitute our entire universe follow strict rules that govern every aspect of their movement and other changes, then at each instant, the state of the entire universe is a necessary outcome of the previous state (combined with those rules). This deterministic necessity, which promises a kind of perfect order that is a dream come true for those who like perfect order, causes trouble for concerns like free will, and leaves no room, it would seem, for contingency. Quantum physicists Flavio Del Santo and Nicolas Gisin, likely inspired by their research into the (purportedly) less deterministic quantum world, hypothesize that classical determinism may be based on an unproven but almost universally held assumption that the properties of the objects of classical physics, particles and such, have in reality infinitely precise values. Of course, our measurements of those values, such as the mass of a molecule, can be only so refined, achieving a certain degree of precision but offering no further information beyond that degree of precision. But this limitation has always been understood as a shortcoming of our measurement tools and procedures, which can never match the infinite precision of the *true* value being measured.

To discard this assumption introduces, or at least makes room for, an indeterminacy that would apply to everything. We (not just physicists, but everyone) are so used to making this positivist assumption about the nature of reality that it is difficult even to conceive of the alternative. What if, propose Del Santo and Gisin, real objects have properties with values that are only finitely determinate, determined up to a degree of precision, but *do not have* more precise values? What if, beyond a certain degree of precision, values just aren't determined in reality, so that reality itself has a limited precision? Borrowing a model of indeterminacy from quantum physics, as exemplified in Erwin Schrödinger's famous thought experiment in feline mortality, the authors posit that measuring a given property will necessarily determine it up to the degree of precision of the measurement. Which means that this hypothesis of a world of indeterminate values is just as plausible, just as consistent with the evidence, as the long-assumed determinacy of the world, because the two metaphysical hypotheses yield exactly the same experimental results.

This proposal about the partial indeterminacy of the whole universe makes an interesting complement to chaos theory, the core premise of which is the idea that some processes are extremely sensitive to tiny variations in initial conditions. That is, extremely small differences in a value—chaos theory was literally born out of a small rounding error—can make big differences in the eventual dynamics of a physical system. So, if those tiny differences of value lack precise values and are determined only to a limited precision, this means, in effect, that the universe gets to make an unconstrained choice, to pick a direction among wildly divergent possibilities. In other words, the theory of finite precision introduces into the universe (and the physics that describes it) a pervasive, radical *contingency*.

The axiom of identity, familiar from grade-school mathematics, provides a formal criterion of positivism. To take as given that a thing is equal to itself, $\alpha = \alpha$, is tantamount to declaring α a posit, the sort of thing that can bear relations of equivalence, that maintains a stable and determinate identity. This might sound obvious: how could a thing fail to be identical to itself? The seeming obviousness of this criterion of positivism, the principle of identity, attests to the mundane familiarity of the positivist outlook, so embedded in our ways of thinking and seeing as to disappear, like reading glasses.

Positivism also institutes a principle of disidentity, which reinforces the autonomy and phallic assertion associated with posits: α is the only thing equal to α; if $\beta = \alpha$, then β is α. Anything that isn't α must have some positive difference from α, and therefore cannot be α. This complementary axiom of positivism demonstrates the subordination of difference to identity. The positivist outlook understands difference only in terms of identity, difference as a selfsame, determinate posit, an existent quality or attribute held by one thing and decidedly not by another. In positivism, even difference, even relation are rendered as static and thing-like.

Hesitating between being a way of encountering the world and being a way that the world presents itself to encounter, positivism is not simply wrong (or wrongheaded); rather, the complaint is that positivist thinking, which favors the clarity of winning and losing over the open-ended negotiation of *multiple perspectives,* leaves no room for other comportments, other ways of being. A world treated as posits offers much of value. It opens the path to numeracy and calculation, discreteness and rule, but also to the liberal individual and her private property; positivism is deeply enmeshed throughout our world and in our ways of being in that world. It is, like the digital, a fundamental way of things, and cannot be just pushed aside by some warmer and fuzzier outlook. Indeed, positivism animates some of our central myths.

Monotheist religion, which recurrently and somewhat defensively asserts the *existence* of the one god as a core tenet, often adopts an implicit positivism as the ontological correlate of this belief in an ultimate existent. One telling episode depicts Moses, anguished about the Israelites' enslavement by the Egyptians, venturing into the desert to look for answers. Finding a burning bush, he converses there with God, who offers reassurance that the people will be freed. Moses will take the news back to the enslaved Israelites, but worries that they will not believe him, and asks God what he should be called, hoping that a name will serve to authenticate the conversation. God replies rather cryptically, "I am that I am" (Exodus 3:14). The structural similarity between this response and the axiom of identity cannot be missed. God avers that he is the first (and final) posit, and after all, to be that which by definition *is* is a pretty solid form of authentication. Philosophers have often recapitulated this self-asserting status of God in various versions of the "ontological argument" for his existence. The argument propounds that existence is not only a characteristic of posits, but is itself a positive characteristic, the sort of characteristic that either *is* or *is not* the case. Moreover, goes the ontological argument, it is God's nature, by definition, to have the characteristic of existence. Thus, God's existence is necessary because existence is included in his essence. (René Descartes's version of this argument extenuates the point somewhat by noting that God is perfect and that whatever exists is more perfect than what does not exist, so that God's definitional perfection implies his actual existence.) Rereading "I am that I am," we understand God to be saying that he is the sort of being whose way of being is first of all to be. Other things can be or not be, but God is that unique being for whom being is integral to his being. To assert a kind of existence that exists out of itself alone and not in relation to anything else is to make the primordial claim of positivism, a sole existent, laid bare, existing out of nothing but its own existence, with no ground, no dependency, no condition on its being.

Descending from the sacred to the mundane, consider the way people talk to the voice assistants of their smartphones and other natural-language-recognition agents. These devices in practice encourage a range of decidedly unnatural vocalizations, such as hyper-enunciation, in which each word must be carefully articulated to ensure the maximum likelihood that the digital device will correctly receive the intended instruction. Speaking to their devices, people typically insert a short gap between each individual word, such that each word becomes a partial instruction unto itself, sonically discrete and distinguished from the words around it. As when listening to native speakers of a language that one has learned only through studying texts, an observer of this "smart" speech quickly recognizes (by contrast) that human language does not ordinarily proceed as a sequence of individually articulated words, but is a flow of generated meaning, a continuum of vowel sounds punctuated by consonants that do not so much discretize those vowels as modulate them at their points of inflection. Individual words typically emerge only through an analysis after the fact, though much written language masks the continuum of speech behind a positivism of words divided discretely by spaces or other indicators. (Classical Greek and Latin writing used *scriptio continua,* writing without spaces between words, but this practice gave way to spaces between words starting in the seventh or eighth century.) Natural Language Processing (NLP) techniques for digital systems have improved dramatically in the last decade or so, and one expects that eventually these systems will accurately recognize sequences of words voiced naturally without hyperenunciation. But the current habit of speaking to one's phone in individual words at this intermediate stage of commercial language recognition is not accidental. Digital approaches to language typically treat words as individual units (and phonemes as individual subunits), attempting to recognize speech as a sequence of words and to capture the meaning of speech by performing calculations on those linear sequences of individual words.

This is not to dismiss the degree to which positivist approaches to language capture linguistic meaning; indeed, much of the meaning of language, including spoken language, can be extracted from a consideration of words in linear sequence. Rather, the point is that this narrow understanding of how language works ignores other important dimensions of meaning-making, and thereby not only misses out on vital aspects of human meaning but also reinforces its limited perspective on language, blotting out the other contributors to meaning and blinding us to their existence. The success of computational linguistics and its practical arm of NLP is hardly total, as digital machines have yet to achieve the language facility of people, but they are successful enough and impressive enough and sufficiently promising of future improvement that many are willing to take the problem of language as virtually solved, which then comes to serve as proof of the validity of the computational method. (See the discussion of GPT-3 [the third iteration of a Generative Pre-trained Transformer system developed by OpenAI] in the concluding chapter of this book.) If computers can (almost) understand and generate language, then their model must be the *right* one, or so goes the implicit argument, and the marginal aspects of language that they ignore must not really matter. Can we discover in human language practices surrounding computation, such as texting or cubicle discourse, evidence of an evolution of language toward its positivist model? Are we remaking language in the computer's image of it?

As the dominant epistemological paradigm, positivism governs not just how we understand the world but also what sorts of claims, arguments, and ideas we count as compelling or legitimate: only an appeal that treats things, qualities, relations, and even differences as posits holds water, eliminating much nuance in favor of the absoluteness of a discrete, self-asserting existent. Theodor Adorno and Max Horkheimer's great work, *The Dialectic of Enlightenment,* takes this positivist epistemology as a central target: "For the Enlightenment, anything which cannot be resolved into numbers, and ultimately into one, is illusion; modern positivism consigns it to poetry" (4–5). This way of thinking brings about its share of antinomies or frustrations. We are fooled into meaningless wars of words by the positivist fallacy. Because everything real, for a positivist, must be an individuated thing, so must consciousness be a thing, and thus its presence or absence is a sensible and unambiguous question, the hinge on which artificial intelligence turns. The soul, as imagined in religiohistorical discourse, must also be a thing, and so it too is there or not, part of the body or separate from it, and these yes-or-no questions diminish the concept of a soul, giving it no room to maneuver and forcing us each to pick a side. Atheists or material monists might object that there is no such thing as a soul, that it is nothing but an invented metaphysical fantasy, but the terms of the argument, conducted under the sway of positivism, take for granted that "it" exists or does not, that to exist is necessarily to be an *it.* Intelligence is an example in which the flaws of the positivist treatment are now well recognized. It's not just that intelligence is measured on a scale rather than a binary yes-or-no, for numerical measure is a favorite positivist technique. It is rather that intelligence is no one thing, no delimited category, and even the recognition of *multiple* intelligences is only a step in the right direction but not fully adequate to the nonpositive reality of the complex that we call intelligence.

Some of these examples highlight that this ontological term, *positivism,* is also all about values. Taking things and ideas as well defined, individuated, and existent is not merely an abstract choice, a question of metaphysics, since these underlying ontological commitments sustain much else about our relationship to the world, including our ways of relating to other people in it. Positivism, which pushes its proponents toward black-and-white ways of seeing, thus also promotes dogmatism. One produces a binary in order to promote one side. Consciousness, souls, and intelligence are first of all values.

The Messy Ambiguity of Earthly Things

The U.S. Supreme Court's recent dramatic reversal of the fifty-year-old federally guaranteed right to terminate a pregnancy resulted from a decades-long campaign on many fronts, in which reasoned debate about the issue was only one battleground, and likely not the most consequential one. It is nevertheless a bitter irony that the heated argument over abortion rights in the United States is the more intractable because of an *agreement* among many on both sides of the issue: the positivist belief that there is a point in the in utero development of a fertilized egg when that egg becomes a person. Of course, the key disagreement arises immediately thereafter: where to locate that developmental milestone. Is it at the moment of fertilization, implantation, when there is a heartbeat, when there is a nervous system that could plausibly support the experience of pain, when the fetus could survive outside of the womb, at birth, or even sometime later? But only a few on either side of the issue consider the possibility that there simply is not a fact of the matter about the start of life or the status of the personhood of a fetus. Even the underlying moral supposition is treated as though it were a matter of fact: both sides typically take for granted that killing an (innocent) person is simply (i.e., positively) wrong, which foundational accord leaves open to debate only the ill-conceived question about the moment at which life (or personhood) begins. It's likely that at least some of the debaters are cynically simplifying their positions, casting their beliefs in positivist terms because positivism is rhetorically powerful, because an emphasis on the ambiguity of concepts like *life* and *justice* does not tend to provoke the zealous commitment among potential adherents that wins political battles. The clarity of positivism, including its moral clarity, tends to trump its lack of nuance, particularly in the domain of politics.

Friedrich Nietzsche diagnosed the problem with this debate long before it became a major voting issue in United States politics. In his short essay "On Truth and Lying in an Extra-Moral Sense," he argues that the positivist bent of our language fools us into believing in the sorts of absolute and definitive things to which that language seems to refer. "That is the situation of all of us with language. When we speak of trees, colors, snow, and flowers, we believe we know something about the things themselves, although what we have are just metaphors of things, which do not correspond at all to the original entities." Having a word for something and enjoying the confidence of knowing how to use that word, we become convinced that the word stakes out a real category of being. That is, we come to believe that there is a fact of the matter about which things are "leaves" and which are not, to borrow Nietzsche's example here. "This gives rise to the idea that besides leaves there is in nature such a thing as the 'leaf,' i.e., an original form according to which all leaves are supposedly woven, sketched, circled off, colored, curled, painted, but by awkward hands, so that not a single specimen turns out correctly and reliably as a true copy of the original form" (249).

Nietzsche thus aims his criticism squarely at Plato. Plato's doctrine of forms holds that earthly things are inexact copies or simulacra of perfect models that exist not on earth, but in the heavens. These models, Platonic "forms" or "ideas," are powerfully positive, not only discrete and autonomous, but so perfect and pure as to transcend earthly experience. But if Plato attempts to safeguard his positivism by locating it in a realm above the messy ambiguity of earthly things, his account of the mundane as simulacral complicates his positivist ontology.

Earthly things participate more or less and imperfectly in their ideal forms, such that every actual leaf is an imperfect version of the idea of leafiness. But "more or less" and imperfect participation muddy the positive ontological assertion of earthly things, clearing a path for a nonpositivist account of the earthly. Thus, Plato both founds philosophical positivism and points down the path of its criticism.

Another great philosophical aphorist, Ludwig Wittgenstein, struggles throughout his work with the lure of positivism, particularly that of language, developing a unique writing (and thinking) style in order to practice a positivism even while refusing its static assertion of a bottom line, a fact of the matter. His great early-career work, the *Tractatus Logico-Philosophicus*, attempts to articulate a world of perfect positivism, a world constituted of an infinity of simple facts, but at the very end of the book, in a much-debated passage, Wittgenstein seems to suggest that the whole enterprise has left us stranded in an austere but ultimately frictionless universe. A world of nothing but posits ends up feeling staid and frozen, lacking the dynamism of the vital questions raised by ambiguity, fuzziness, and weirdness.

Wittgenstein's later work exercises this same method—building systems of thought in language only to undermine them through interrogation—invoking a persistently hypothetical mood through both grammar and semantic content. As though troubled by an inability to generate solid conviction, he ponders in his writing numerous, and at times bizarre, hypotheses, situations, and claims, examining the implications of each, but usually refusing to arrive at a definitive and final conclusion. It is a method of questing or questioning that increases one's familiarity with the relevant philosophical territory, building up a nonpositive know-*how*, rather than a more assertive know-*that*. A popular reduction of Wittgenstein's views on language advances the cliché that "meaning is use," and while any such formula already says both too much (its very form exemplifies a hardened posit) and too little (its generality floats away from all that is specific about language), it nevertheless offers some sense of how a Wittgensteinian take on language would refuse to grasp meaning by way of purportedly true claims and would instead locate meaning in the living dynamism of speech and writing, language in use.

Echoing Plato, though perhaps unwittingly, neither the pro-life nor the pro-choice proponents choose to read Nietzsche and Wittgenstein, which is hardly surprising, for too much is at stake. Admitting the fallibility of the positivist position would topple a domino at the head of a chain that snakes through sacred territory. It is, after all, the monotheist god who guarantees, in a world that would otherwise refuse any guarantee, the sharp borders of our concepts, as well as the reliability of language in its alignment with the real. God fixes things as equal to themselves, makes the world true (such that we each possess a "beautiful soul," according to Descartes, which can reliably trust whatever in good faith it observes), and finally gives meaningful heft to moral posits, such as the interdiction against murder of innocents, in part by promising a postmortem justice, a day of reckoning. What document advances a positivism more resolute than the Ten Commandments?

Some varieties of postmodernism refuse positivism and its absolutes. Much twentieth-century European thought seeks alternatives to the epistemology of it-is-or-it-isn't that characterizes positivism. One might read these nonpositivist approaches as more nuanced or richer, but by the same token, postmodern thought opens itself to charges of fuzziness or imprecision. To a positivist, the rejection of the clarity of yes-or-no leaves no alternative, and thus appears as a form of nihilism, a descent into the senseless: it is understandable to claim that the soul does not exist, and it is understandable to claim that it does, but to suggest that the question itself is misguided, that souls are not the sorts of things that do or don't exist, is to embrace a confusion that ultimately makes meaning impossible.

The argument herein is that the digital benefits from the prevailing positivism, for it is a technology of positivism par excellence, but also that the digital reinforces positivist thinking, training its users and the culture that accommodates it to admit as meaningful or legitimate only what can be posited. Only a positivist form of information can enter the digital machine, become data or instruction in the computer. In the most concrete sense, the binary code, which is the digital's processing language, is already in total conformity with a positivist outlook. The binary code captures everything, both process and product, as sequences of 0s and 1s, discrete, self-identical, unambiguous numerical values, that might be the apotheosis of positivism, Platonic forms instantiated here before us on earth. Bits are a kind of ultimatum of positivism, a first principle. The founding gesture of digital technology is to count the bit as a thing, to treat it as an independent, empty posit, a *yes* or a *no* that waits for a question to give it meaning, and so can be given whatever meaning one desires by posing the appropriate question.

Digitizing Gender

Gender is beautiful. Even its arguably intractable binarity is beautiful. *Vive la différence!* That binarity, a distillate from the enormous complexity of a push–pull, mostly demonstrates the absurdity of strict category when it comes to human difference. The point is that gender is both a binary and an inexhaustibly rich field of (in)determination. The binary, in this case, derives somewhat fitfully from that rich field, but it is not *simply* imposed on that field, as though the gender binary were only a vehicle for the differential distribution of power, a division that sows division, to tie up difference, bind it to category in order (ironically) to neuter it. Though it is true that binarized gender does neuter productive difference with tragic consequences for some individuals and for the socius generally, it is at least as true that binary gender engages a great deal of productive difference, both within its binary and without. Gender oozes with difference, between its binary categories, within each category, and in defiance of those categories. And we are discovering (perhaps always discovering) that many vectors, many conformities with and many transgressions through that binary, are possible. (So that the passage away from the binarity is not an opposition to it, which would only construct a new binary.)

But that is precisely what is beautiful about the encounter between gender and binarity. The binary does not dominate gender, which remains productive around and against that binarity. The digital, inexorably consigned to its own binary, can never approach gender, never capture but its shadow, its remainder, left behind after it exhausts productive difference.

Gender's productivity complicates the identity categories built around it, such that its politics is pulled in (at least) two directions. The positivist allure of category, seemingly a prerequisite of a politics of recognition, finds a ready grammar in our current discourse, and opens the possibility of alliances with many other categories, intersections, as they are often called. But the opposition implied by the category, a binary opposition that belies the productive difference that gave rise to it, is in tension with gender's generativity, which produces both its binary and its departure from that binary.

Rationalism

In a nutshell, rationalism is the conviction that the world makes sense. Its most prominent precept is the *principle of sufficient reason,* closely associated with the Enlightenment philosophy of Gottfried Wilhelm Leibniz, which holds that there is a reason or explanation for every event and every thing. From one perspective, this is an immensely optimistic outlook, for the world appears to be full of arbitrariness and accident, making it seem unlikely at first glance that there is always a sense to be made. A guarantee of sense would be deeply reassuring for many, as, faced with mortal threats and gross injustice, one could take solace in the knowledge that there is an ultimate order, that everything fits together in some grand vision. If there is a reason for everything, if everything that happens happens for a reason, then it is a short step from rationalism to universal necessity. That is, if everything happens for a reason, then those reasons explain in aggregate why things are the way they are, and this would seem to weigh against things being some other way, hence a world of necessity.

Though plenty of explanations appear to be unavailable or at least hard to find, we tend nevertheless to adopt a tacit rationalism, to believe that any given situation can be explained, even if the explanation is not readily available. Typically, we reject the notion of an event or a thing that simply has no explanation, that comes about entirely accidentally, and so cannot be understood as a result of any cause. We are inclined to think that, at some level, there is a reason that any given event unspools the way it does. Rationalism thus works well alongside a secular adherence to material monism, the belief that all there is is physical substance and that everything we see and do is made of that physical substance. Material monism would plausibly offer an explanation for every thing and every event, an explanation in terms of the laws of physics and the dynamisms that they imply. On this view, even thought comes about as a result of neurochemical processes, which themselves reduce to causally determined microphysical interactions. And complicated events, like professional sports competitions or the results of a national election, come about as the outcomes of a huge number of neurochemical (and other) events, all of which boil down to physics.

Even if it leaves no room for real mystery, for what might be essentially unknowable, rationalism offers a significant comfort, for it assures those under its sway that not only can everything be explained, but with the right intervention, most things can be duly influenced. It is closely tied to a modernist fantasy of control. If everything makes sense, if the world is ordered and reasonable, then by applying the appropriate measures, one can achieve the desired result. In this sense, rationalism is the perfect complement to instrumentalism, for it justifies instrumentalism by affirming the cause-and-effect or means–ends reasoning on which instrumentalism can reliably proceed.

Digital mechanism ensures that causes are inextricably linked to their effects in the digital, that the digital proceeds deterministically and inevitably to produce the precise results that its instructions mandate, making the digital a microcosm of thoroughgoing rationalism. Unlike material technologies that admit accident, happenstance, and unpredictability, digital technologies, as we will see, reduce or eliminate accident to forge an immaculate connection between instruction and execution, between cause and effect. But, even aside from this seamless joinder of cause and effect in the digital, it is a crucible of rationalism because the nature of digital operation is that every digital event, every step of digital process, always has a reason. The digital unfailingly makes sense— albeit a stunted and formalistic sense—because it everywhere and always operates according to a simple binary logic. Even the most unexpected outcome can be reliably explained by the simple but very numerous operations of binary logic that produced it, such that any digital surprise attests only to the relatively weak (or slow) capacity of humans to perform the calculations that would have yielded the correct expectation.

The formal sense that the digital makes is therefore not usually very satisfying. To account for an aggressive forum post or the relaxing effect of a digital ASMR (autonomous sensory meridian response) video by appealing to a sequence of billions of binary calculations feels like a parody of sense-making, leaving the inquirer with no "real" explanation. It is akin to explaining a person's motives for choosing a career path by listing the neurochemical interactions in her brain. Though it may still fall short of adequate explanation, digital rationalism does offer more than just a list of binary operations; each operation is one in a sequence of operations, and that sequence has been designed, as hardware or software, to accomplish specific tasks, to make available certain possibilities of the digital machine. *Design* is a key term in this instance: that a sequence of operations has been designed indicates that it has been inscribed in hardware or software with deliberation, in order to carry out some digital process, trivial or complex. The author (programmer or engineer) of a subroutine (sequence of operations) may not know what greater ends her contribution will serve, but the subroutine itself is not open-ended. Each subroutine accomplishes something (or somethings) specific, and each answers to its designer's intention. Just as the digital is an apotheosis of positivism, so this thorough intentionality of the digital machine redoubles the formal rationalism that follows from its slavish mechanism. Every digital action has a reason, and that reason is not capricious or imprecise, for it can be traced to a determinate intention of its authors. The digital works entirely by design.

Science and Religion

Both scientism and some religion are built on top of a rationalist premise. Though the Lord might work in mysterious ways, this reassurance is offered precisely to ward off the threat of sense-lessness or unreason. The Lord has his reasons, and though they may not be accessible to us mortals, we should be satis-fied knowing that his will serves as the one real meaning of the universe, which gives everything its sense or reason. But a faith in science addresses a similar worry about the threat of unrea-son. Though many scientists likely defer when asked whether we can discover ultimate reasons through science, it neverthe-less provides what many people regard as the most authorita-tive reasons, the "real" explanation of how things work or why things happen the way they do. Note how intelligent design conjoins the two discourses of religion and science, suggesting, in effect, that the order revealed by science has its reason in an intelligent designer. (But, as with many arguments about god's existence, one is tempted to ask in this case what the postulate of a designer adds to intelligent design. Why is it any more sat-isfying of an explanation to locate the source of the universe's order in a self-same will, a universal subject, rather than attribut-ing that order to the universe itself? How does it explain order to cite its designer, when one can simply acknowledge order as part of how the universe works? As the argument for intelli-gent design sets forth, if the universe did not include order, we would not likely be in it, for how could something so negatively entropic arise in the absence of stable order?)

It is telling that these often irreconcilable discourses, science and religion, both answer to the same desire. And what it tells is just how ardent this desire is for so many people. A world with-out an ultimate sense, a world in which things might not finally make sense, is a world that many simply cannot tolerate, so they embrace the nominal sense offered by certain forms of religious belief and certain uncritical attitudes toward science. There are also plenty of adherents of religion and of science who adopt a greater humility, recognizing that even faith leaves some questions unanswerable. Perhaps this abandonment of rationalism provides a platform that is ultimately more durable because less shaky, though the predominance of rationalism as a discursive paradigm means that its abjection is not likely to win a lot of arguments.

Digital rationalism, wherein every digital process makes a minimal or formal sense based on binary calculation, relates closely to the digital's stunning efficacy. Digital technologies operate by breaking down tasks and information into small parts, and ultimately into bits. All of the information in a digital device and all of the operations of a digital machine can be, and in practice are, broken down into pieces tantamount to yes-or-no questions, and the universality of this resolution into elemental bits is the guarantor of the sense of every digital process. Imagine giving a person instructions in that same format; a person might have thousands or millions of possible actions available to them at any given moment, and you convey the correct next move exclusively through a sequence of yes-or-no responses. "Does the next move involve using my body?" *Yes.* "Does it involve one of my limbs?" *No.* "Does it involve my torso?" *Yes.* "Is it bending?" *No.* "Is it twisting?" *Yes.* And so on. This doesn't seem to be how people think or act, but this kind of action made of tiny individuated microactions is the only way that computers get things done, a game of "twenty questions" but with way more questions. And it illustrates why the mind-bending speed of computers, the billions of operations they can carry out each second, is essential to their success: when each action is so insignificant as barely to make any difference, or makes a difference only on the order of a single bit, nothing will get accomplished until many many such actions are sequenced. The computer's speed makes up for its microscopism, its way of getting things done only in the tiniest incremental steps.

This same extreme of tiny steps, the breakdown of every action into the smallest bits of progress, also gives the digital much of its nearly universal reach. By breaking things down into such tiny steps and such tiny bits of information, the digital device possesses a starkly simple basic vocabulary with which one can say almost anything, given enough of these tiny signifiers. An analogy can illustrate this power of the elemental: using a tool that makes centimeter-long line segments in different colors, one can draw many things, but using a tool that makes tiny dots in different colors, one can draw almost anything. It takes a lot more dots, a lot more operations of the dot tool, to make a drawing that could also be made with the line tool, but the additional flexibility of the dot tool provides a significant advantage, especially if you can apply the dots so quickly and repeatedly that you don't really care how many you have to use.

This strategy of breaking down tasks into discrete micro-operations, the universal strategy of digital technologies, is a rationalist, positivist fantasy. It generates a world of individuated posits, manipulable through elemental actions that establish a complete vocabulary of all action, a basis for every possible action. One consequence of the achievement of this fantasy is the flexibility of the system once atomized. One can even achieve radically opposed ends using the same set of means just by sequencing them or distributing them differently. A value or meaning advanced by the overall task might not be discoverable in any particular microact included as part of that task. Breaking things down into atoms thus has the potential to strip those atoms of their significance, leaving them agnostic and available without prejudice, for any purpose.

Instrumentalism

Agnostic or inert, the digital makes itself wholly available for instrumental appropriation. Instrumentalism treats the world in terms of means and ends. An instrumental outlook takes things, often including people, as instruments by which to bring about some desired state of affairs. It is thus frequently lamented as a dehumanizing defect of our technocapitalist era, which encourages the mercenary treatment of everyone and everything. Our lives are regulated according to prescribed ends. A meaningful existence is defined by achievements, usually measurable, and we are encouraged to invent goals and calculate the means of their accomplishment. One exercises *in order to* live longer, or lose weight, or win a competition, or improve one's attractiveness, or even just because it feels good, which is a well-rehearsed end. One works a job in order to pay the mortgage, or help others, or climb the ladder, or avoid idleness, which latter would be an insult to instrumentalism. A zealous instrumentalism rejects both the community of people and the community of nature, since value, of persons as of things, derives from individual utility and not from the more nebulous context in which actions repercuss indefinitely. Ties to a community are at best superfluous to the utility of an individual and tend to interfere with its full availability for instrumentalization. When an animal, mineral, or vegetable reveals its embeddedness in an ecosystem, it becomes much less exploitable, as would-be users must confront the scale of disruption that its use would engender. Inasmuch as relations impede instrumentalization, instrumentalism functions best alongside positivism, as the autonomy of a posit renders secondary all relations, and thus removes from consideration concerns about unintended consequences and other forms of resistance. A posit, which is nothing but a bald assertion of individual being, can give itself entirely to utility, unencumbered by other ties and other concerns.

Technology in general is evaluated almost exclusively in terms of its instrumental value, an observation that opens Heidegger's discussion of technology: technology is popularly understood firstly and mostly as a means to human ends. As with positivism and rationalism, the digital pushes to an extreme the association between technology and instrumentalism, for digital technology is wholly designed, each of its actions specifically constructed and executed to realize some envisioned end. Each operation, from the minutiae of binary arithmetic to the sequential processing of billions of binary calculations in a complex algorithm, takes place for an intended reason, serving as means to an end.

Given the prevailing instrumentalism of our moment, the digital's heralded utility lends its technologies a seemingly irrefutable superiority. It gets things done so well that there remains little room to question whether to adopt it when available. A single device can accomplish an astounding multitude of tasks, occupying increasingly small spaces, consuming minimal resources, and operating with blinding speed and unfailing accuracy, the digital as an archetype of efficiency. "Going digital" has become the obvious choice, making other media and other means feel nostalgic or antiquated, quixotic or luddite. Instrumental efficacy wins out over almost any other concern; results are presumed to be the foremost decision criterion, trumping other relevant principles.

Designed from start to finish, digital technologies are structured by the ends at which they aim, but that aim is unusually true, as the digital realizes its end with near infallibility. In the digital, there is no *try*, there is only *do*. That is, the digital does precisely what it is supposed to do, where the relevant supposition exists not in the mind of the user or the author of the subroutine, but in the expressed content of that desire inscribed in a digital form. A program may or may not accomplish what the programmer had hoped or intended, but it will certainly accomplish what the programmer, in her encoded instruction, actually commanded it to do. And that without fail.

A system of means and ends reveals the mutual proximity of rationalism and instrumentalism. These ideological orientations are two sides of the same coin, instrumentalism leaning toward possibilities of action and rationalism underwriting the ontological structure that makes action possible. That which has a cause, a reason why it is the way it is, therefore also has a potential site of intervention, a way of affecting the outcome. Cause and effect map conveniently onto means and ends. And to think or act instrumentally, in terms of means and ends, is to locate a reason for every outcome, to institute a rationalism that reads a reason from its goal or end.

Digital machines are entirely goal-oriented. Each user action at the machine is undertaken in anticipation of accomplishing some goal, often on the way toward a grander, though still digital, ambition. One clicks the mouse or presses a key in order to accomplish something, however slight. Of course, one can interact with these devices haphazardly or idly, mashing keys at random, or to hear the sound of the tapping, or to relieve the stress of maintaining such a rigidly goal-oriented bearing while using the machine. But even the most undirected activity at a digital machine requires at least a minimum of goal-setting, for the machine *responds only to commands,* and those commands have predetermined (and typically predictable) outcomes. Every interaction with a digital machine takes the form of a command. A database query might appear on its face as having the form of a question, such as "How many cities in North America have a population greater than one million people?" But this yields a response only inasmuch as the machine takes it as a command (or sequence of commands): "Search this database for

North American cities with population greater than one million, then count the number of results, then output the result of that count." Every click of the mouse tells the computer to do something, run some algorithm, execute some procedure. Every press of a key likewise functions as a command. Commands are the only mode of input the computer can accept.

Automated processes can derive commands from an ambient or passive environment, leaving those commands implicit. Watches and phones track and store the user's location at regular intervals without being told to do so each time. Once properly set up, smarthome software turns lights and heating on and off with no explicit instruction from the residents, and that same software can even adjust its parameters via algorithmic guidance, again without any ongoing human intervention. Our online actions are notoriously surveilled without our explicit permission: clicks and keystrokes generate data that feed algorithms that perform all sorts of responsive actions, from serving targeted advertising to entering user names and associated data into a database for purposes to be decided only later. Notably, these examples do not illustrate computation in the absence of commands, but only the way in which automation operates through deferred commands, commands that are given in advance of their execution. And in such cases, even the data collected invisibly operate as commands to the digital machine, prompting the machine to store the data for later access or triggering the execution of some context-dependent algorithm. We don't always know the commands we are inputting, and we don't always know when we are issuing commands, but for the digital machine, nothing happens without it being commanded.

As such, the digital will always be a means but never an end in itself; the digital is inert or inactive on its own, waiting for a command in one form or another, and its operative mode is inherently reactive. Put otherwise, the digital is completely submissive; it has no will, so that it can carry out the user's will, or rather, the will expressed in her commands. The machine acts only when (and how) it is instructed to do so, and any instruction is tied to some expected outcome. This claim of complete submission should be carefully distinguished from the clearly false claim that the digital always does what the user desires. The digital may save labor in many ways, but anyone who has used a digital machine knows that it institutes its own regime of labor, which is precisely the labor of translating one's desire into a positivist, rationalist, instrumentalist form, into the form of information. A recurring fantasy of the digital is to eliminate this labor of instruction, such that the digital would not only execute one's commands, but somehow anticipate them without the user having to instruct it. Apple cofounder Steve Wozniak describes this as an unrealized design goal of the Macintosh computer: "Early on with the first Apples, we had these dreams that the computer would let you know what you wanted to do." But, while digital technologies attempt to simulate this takeover of one's will, say by guessing that the user will wish to enter a calendar appointment when she receives an email with data appearing to describe a future event, still the rule of commands is unwavering and the digital can demonstrate this simulated intuition only if it has been given explicit instructions for recognizing appointment-related data and what to do when it detects those data. The user's will is not really being anticipated, which is why this "smart" capacity of the machine must still ask the user whether to put the appointment in the calendar or to skip the automated data entry this time around. There is good reason to think that Wozniak's dream can never be realized, that the labor of instruction is irreducible.[4]

Standardization, which is rampant in digital design, safeguards digital availability or instrumentalizability. A specification, or "spec," is an ends-oriented presentation of standards. The spec specifies the ends, outputs that must result from particular inputs, but thereby leaves unspecified the means by which those ends must be achieved. There is more than one way to translate a command written in source code into a sequence of binary calculations that can be directly processed by the digital machine, but the spec ensures that the outcome will be identical regardless of just how this translation is undertaken. Standards thereby close the gap between input command and execution, ensuring that nothing irregular or unanticipated can happen in the opaque innards of the machine, where most of the processing happens beyond the user's ken.

Digital devices are especially instrumentalizable, more cooperative than mechanical tools, in part because they operate in a world apart from the actual. As described in chapter 4, digital calculations unfold in an ideal space, separated from the resistant materiality that might refuse to proceed as instructed. Will-less and exclusively obeisant, the digital is confined to its role as instrument, a pure means whose only ends are those inscribed from without.

Standards, though not unique to digital technologies, are uniquely effective there because the digital is imperturbable: a digital standard allows no deviancy, cannot be shaken by accident. Other technologies are exposed to the unanticipated, to a world of factors beyond their reach, whereas the digital erects its own world, a world made of information, and the passages between its world and ours are strictly regulated. Erecting its own world in which it is also the gatekeeper, the digital establishes dominion there, a world of digital rules, a world where all that happens results from the commands that govern it. It is therefore supremely available, the ultimate instrument, for one need only give it the correct command and it can do nothing but follow.

In his 2011 *You Are Not a Gadget,* Jaron Lanier exposes an unfortunate effect of the digital's reliance on unwavering standards, arguing that those standards can outlive their utility, imposing limitations or demanding accommodation even when the tasks to be performed have shifted in some way and no longer match well with previously established norms. He calls this inflexible rigidification "lock-in." The standardization of any format or operation constrains future changes, since some changes would require reconceiving the standard itself, but because the standard is prerequisite for the many powerful software operations that rely on it, changing a widespread standard is a socially and technologically overwhelming task. Among Lanier's many examples of lock-in, the reliance on the file as a basic way of organizing digital data stores may be the most deeply embedded and intractable. Today's operating systems take for granted that data will be stored in files, and it likely rarely occurs to anyone that there might be a better or different way to organize data. The core argument of this book mirrors to some extent Lanier's logic of lock-in, though on a grander scale: the claim here is that the commitment to discrete code, irrespective of the content of that code, already leaves out certain possibilities of creativity and representation, possibilities that cannot be reinjected into the digitally enabled processes that exclude them. The result is that those possibilities come to be forgotten or ignored as unimportant, altering the entire landscape of creativity in our world.

Confined to its own world and enslaved by the rules that determine its process, the digital cannot err. Of course there can be error in the digital, but there is no digital error. That is, as a *digital* process, the digital infallibly generates the results of its binary calculations; the laws of physics guarantee that those calculations proceed exactly as they should. An error could be a matter only of giving the digital the wrong instruction, which it will nevertheless unfailingly and blithely follow. In this narrow sense, the digital works *perfectly,* a perfection that marks its unique status among technologies, its unlimited utility, but also points to its insurmountable shortcoming in relation to the actual. Its perfection sets the digital apart, founding the *digital reality* where its reign is absolute, but also denying it a vitally enriching participation in our reality. How does it achieve this remarkable perfection?

Digital Game Worlds

A digital game includes a completely designed world. Whatever actions the player can take, those actions have been planned, enabled, prescribed by the game makers. Then again, gameplay possibilities occasionally arise that are, to a degree, accidental affordances of a game. The interaction of sets of rules written in digital code that were not carefully checked against each other may make it possible to walk through a wall at a particular spot or repeatedly gather loot from the same chest or dispatch the powerful boss in a single move unanticipated and overlooked by the game's designers. Such accidents are often shored up in updates to the game, as the aim of the designers is generally to keep players playing within the intended rules, and the rules of the game cannot be strictly disentangled from the rules of the code that make the game run on a digital machine. Sometimes a *glitch* or *exploit* is allowed to remain if it is sufficiently amusing, or challenging to execute, or difficult to remove, or benign in relation to the balance of success and failure, the overall difficulty of the game.

The totally programmed nature of digital games, which are an expository model of how all software operates, seems as though it ought to frustrate the sense of freedom that is, one would think, part of the pleasure of gameplay. A player in a game might enjoy all kinds of programmed capacities accompanied by a sense of freedom from constraint: to sneak past a nearby armed guard by hiding in a shadow until the guard looks the other way; to fire two large, heavy machine guns, one in each hand, and still be able to reload in fractions of a second; to recover instantly from a spiked baseball bat to the torso; but that same player cannot scratch her nose or peek around the corner of a doorway, because those affordances simply were not programmed into the game. Gameplay is precisely the willing accession to prescribed limitations; figuring out how to play successfully *within the rules* is the central task of gameplay. Moreover, as Christopher Douglas points out, these hard edges that determine what a player can't do (and allow what she can) contribute to the pleasure of the game, for they reveal that the entire game world has been designed with the player (or her in-game capacities) in mind: ledges are placed at exactly the height that the avatar can jump, ammunition is lying in unlikely places on the ground at just those points where the player's reserves are almost empty, and so on. There is a comfort—Douglas calls it "existential soothing"—in playing in a world designed around oneself, a narcissism born of the sense of one's own perfect accord with the world one (or one's avatar) playfully inhabits.

This quality of a world designed around the protagonist is not original to games but characterizes most narrative more generally. Greek tragedies and traditional fairy tales exhibit a certain economy of actions and objects, a moral and metaphysical economy, in which every object, every character, every action has its proper place, fitting together like a puzzle. It may be constitutive of satisfying narrative that its world seems to make sense, to suit the actors within it, even if that sense is tragic or inadequate or existentially challenging. That is, a character might undergo a series of unfortunate events, but it is in that character's unhappy outcomes that we feel something fits as it ought.

Think about the consequences of the rule-bound, what-you-code-is-what-you-get nature of the game. It means that the capacities of the game characters, the avatars (or the possibilities of acting in the game, for games that don't have or don't much use avatars), are likely determined by the ludic logic that defines the core of the game. It is the rules of the game, the invention or conception of certain puzzles, the creation of a certain (simulated) physical layout, and the scarcity and occasional provision of certain resources that end up determining what the player can do. The player's abilities are put in place *in order to* do what needs doing in the game. The avatar (or controlling player) is not first of all endowed with plenary abilities that then come into play in the game. Rather, the player's affordances are designed for the game and make possible not so much freedom of action as correct or incorrect deployment. The instrumentality of game design intrudes uncomfortably into the creative freedom to which games appear to offer access.

Games that include multiple players in the same virtual environment, especially when those players are afforded unrestricted communication through text or speech, would seem to open an intermediate space, one that admits a greater freedom that is less dependent on the game designers' preconceptions. Possibilities for playing the game remain constrained by the game's rules and the code that both delimits and allows what the player can do, but interaction with other players can establish a space for more inventive gaming, ways of making meaning within the game that are validated by the shared experience of the game space and not necessarily determined by the designed world of the game. What might appear in a single-player experience as a clumsy and pointless movement of one's avatar can come to signify a victory dance when there are other players to witness it. But even if expressive possibilities extend the territory of play, the game's rules, as enforced by the software itself, essentially define the material of expression, not the significance of action in the game but the actions that can be become significant there.

To use a computer is to instruct it, to tell it what to do. Digital technologies give themselves over to the user's desire, they become the user's instruments. Otherwise without resistance, they insist only on a dress code: any entrant to their world, any desire invested in a digital machine, must wrap itself in the garb of positivism, rationalism, and instrumentalism, for the digital can work with nothing else. The reward for conforming one's desire to a digital-ready form is that the digital not only submits to one's will, but becomes an extension of it. The digital, which does only what is properly instructed, carries on the desire of the user, executing that desire even in her absence. We engage with the machine in order to execute, perpetuate, and propagate a will to an end, desire as posit.

The computer has burnished its image by perching atop this ideological ruse, pretending to be a value-free or neutral technology that simply and unproblematically works extremely well. Effacing (by naturalizing) those conditions that motivate its wide and enthusiastic adoption, the digital denies an allegiance to any value propositions, as though efficacy (for example) were a plain and universal truth, rather than a relative and contextual judgment. (To judge something efficacious is to have decided in advance what one hoped to accomplish and possibly also how.)

The connection between the digital and the commitments of *digital ideology* is fairly easy to establish: as a technology, the digital races along according to its simple rules, manifesting a mindless, mechanical consistency, which makes generalizations about it quite reliable. The greater challenge, as is typically the case with ideological critique, is to show that this is indeed ideology, that there are other truly different ways of seeing the world, such that digital ideology, inasmuch as it makes an exclusive claim on our thoughts and actions, is consequential and even problematic. We may be so acculturated to this Enlightenment legacy that we hardly recognize alternatives. When we don't really even conceive of other ways of seeing, the claim that the digital discards those alternatives packs little punch, since we don't know what we might be missing, or even that there is anything to miss.

Lest it be mistaken for a technological determinism, the claim is not that the digital (or its associated technologies) is an inevitable outcome of Enlightenment rationality. Many historical accidents contributed to the coalescence in the mid-twentieth century of technologies and methodologies that became the digital; further, rationalist positivism is not the only Enlightenment ideology, nor the only influential ideology in the intervening four centuries of multicultural development. Rather, the claim is the gentler proposition that digital technology, by virtue of its underlying principles, aligns all too well with the values of positivism, instrumentalism, and rationalism, values that long preceded digital technology, and that this alignment accounts for its prevalence around the world today. It's not just that the digital is effective at getting things done, but that the very idea of getting things done and the very ways in which the digital operates already accord with the dominant epistemology. Moreover, perfectly aligned with those prevailing and increasingly pervasive ways of seeing and being, the digital propagates those ideologies to squeeze out other values and other ways, pointing toward a hegemony of digital being.

◀ 3 ▶

Ontology and Contingency

We inhabit a world of contingency. Without accident, without the unpredictable, the world would be stuck, static, and staid. Yet digital technologies are designed to elide contingency; the digital world, apart from and a part of our own, does its utmost to rule out the contingent. This exclusion makes possible a maximal, even perfect, efficiency, but also interrupts the digital's relationship with those foundations of a dynamic universe that grow out of contingency: meaning, value, change, and the creative. What is contingency, that its erosion in the digital produces such a pronounced anemia but seems to pass almost without notice? If contingency can claim responsibility for the most essential aspects of ourselves and our worlds, how can the digital evade its ubiquitous significance, and how does the digital even exist without it?

Contingency is a term taken from the philosophical lexicon. In ordinary usage, the contingent (from *con* + *tangere,* to touch together) is something that depends on something else. An event, such as a backyard barbeque, might depend on something else, such as cooperative weather. And we say, "We're having a barbeque, *contingent on the weather,*" or something like that. If for some reason it turns out that we must hold the barbeque regardless of the circumstances, maybe because it offers the one chance to say goodbye to a rarely seen friend, then we might say, "The barbeque is necessary, *whatever the weather,*" meaning that it isn't contingent, it doesn't depend on anything but will happen in any case. The philosophical version of contingency means more or less the same thing but places particular emphasis on this contrast between contingency and necessity. In philosophical discourse, what is contingent is what is not necessary (but also not impossible), what might *or* might not be the case; philosophically, the contingent is that which *depends,* but possibly on nothing in particular. More dramatically, contingency is the condition of the unconditioned; freed from necessity, it inhabits an openness or indeterminacy, respecting no rule, owing no allegiance. As this moment of potential, an injection of dynamism into the passage of being, contingency's ontological import comes to the fore: ontologically, contingency is the locus of freedom.

The ensuing discussion fills out this notion of contingency but does not claim much originality in this invention. Contingency, as presented here, is an idea absorbed from a number of different philosophers and philosophies, and then regurgitated as the founding principle of a speculative and figurative metaphysics. Though the ideas presented here do not pretend any great fidelity to the philosophies from which they are borrowed, they owe their greatest debt to Gilles Deleuze. Readers familiar with his ideas will discover much synonymity here. Martin Heidegger, too, exerts considerable influence over the thoughts in this chapter, but would surely reject the proposed metaphysics as poorly grounded and as violating the phenomenological basis on which he felt any metaphysics must proceed. In any case, it is something of a conceit to name this hodgepodge of ideas "contingency," as that name appears only occasionally and not foundationally in the original sources. Nevertheless, it is hoped that contingency and the metaphysics constructed around it will exhibit a sufficient internal consistency and compelling apposition, by which readers can verify its legitimacy in their own intuitions.

Contingency and positivism are opposed, and the adoption of positivism is only one of the barriers the digital erects to ward off contingency. Indeed, contingency disrupts all three ideological commitments of the digital. It undoes *positivism* by refuting the strict identity of a thing with itself; it denies *rationalism* by asserting a spontaneous sense that pierces the limits of reason; and it foils *instrumentalism* by destabilizing the relationship between cause and effect. The rejection of contingency thus serves to conserve positivism, rationalism, and instrumentalism, and so perfects the digital, sustaining its reliability and consistency, even its speed and lightness. Without contingency, a digital technology can proceed unperturbed and unhampered, for there is nothing, no surprise, no resistance, to impede its progress. Contingency's banishment is the signal gesture of the digital, its most fantastic and consequential innovation. But it is also the hardest limit, the defining deficit of the digital, ensuring its poverty, its formal senselessness, its flattening of value, its endless spinning in place.

Our most typical understandings of the world do not recognize the import of contingency, for we are beguiled by those same ideological stalwarts that so favor the digital and embrace its entry into every domain of human endeavor. To buy into positivism, rationalism, and instrumentalism is to reject contingency as value and as fact. Because it is antipositivist, contingency does not present itself as a posit; its way of being is not to be *there,* where one might point. It does not set itself forth, stand before our gaze, assert its presence. This makes it hard to see, hard to articulate, hard to identify, hard to describe. The very idea of understanding has a positivist bent, such that it might make no sense to suggest that one could understand what is not a posit. Rather than admitting direct scrutiny, contingency's influence is felt at the margins of things, at the thin edges of sensation, in the fluctuations of being that surround every particle in the universe invisibly but consequentially. Contingency is a *way,* an event, a happening, but not an object or a thing.

The lack of condition or freedom of contingency motivates its expansive definition in this book. As used here, contingency retains its essential opposition to necessity but further extends over a range of related ideas to demonstrate its core significance in disparate domains. Philosophers, among others, have given various names to the creative principle—difference, becoming, freedom, uncertainty, accident, noise, virtuality, groundlessness, faith—and though the relationships among these different ideas are complex and they are not always compatible, contingency overlaps with each to draw a ragged and unconventional history of ideas. For instance, as already referenced, contingency in one of its facets is a phenomenon of the margins, the thin edge of an object or the long tail of an event, and chaos theory studies those situations where the center depends on the margins, where the main idea exhibits a surprising sensitivity to minute differences in initial conditions. (Deleuze's analysis of singular points in calculus, at the outset of chapter 4 of *Difference and Repetition,* casts such points in a similar role.) Of particular relevance to the analysis of the digital in relation to contingency is its refusal of transcendent rule, the potential for any rule to be trumped, recast, delimited, reinterpreted, transgressed, and so on. Immanuel Kant's insistence on the spontaneity of reason recognizes this same potential, that sense respects no final rule but serves always as its own judge and jury. (Then again, Kant's categories of the understanding stand in his philosophy as inviolable rules, for nothing can be thought outside of those categories.) Still another facet of contingency shows its affinity with indeterminacy or undecidability, contingency as the nonidentity of a thing; shot through with contingency, a thing is always becoming something else. And in its clearest opposition to necessity, contingency arises as creativity or freedom, an unboundedness, the preservation of choice, free will, openness, a potentiality that has not been tamed or corralled. Deleuze discusses this potential as the "virtual," a realm of decision as yet unmade, where what *might be* is assembled.

Philosophical Antecedents

To mark an ontological distinction between digital and actual, as this book does, risks an immediate confusion. The digital, or at least its associated technologies, *is* actual, and so must surely share the actual's way of being. It's true that, as material objects, computers are exposed to contingency, as is all material. But as *digital* machines, computers also occupy another world, with different rules, a different logic. Superficial evidence of that difference requires only a glance at a running computer. Unlike any actual object, for example, an object on the computer screen can be readily transformed in an instant into just about anything else. It can get bigger or smaller, slow down or speed up, change any of its parts without affecting the whole, and indeed undergo any arbitrary transformation with practically no residue, no resistance, no effort. There can be digital mechanisms to, for instance, limit access to data. The data themselves are not *sticky*, for it is a basic principle of digital operation that the 0s and 1s that constitute objects and events can be changed to other sequences of 0s and 1s. Indeed, computer operation is made of such changes of 0s and 1s. By contrast, the actual world typically demonstrates an ontic inertia; human-scale objects tend to change only slowly and according to some reason. But in the digital, anything can become anything and there is minimal cost associated with that becoming. This apparent weightlessness of the digital, the lack of consequence or ground, the way in which things are not anchored to themselves, is a big part of what we usually mean by the *virtual*. Because the digital world is (in this sense) virtual rather than actual, it makes available all kinds of ready possibilities that would be hard to reach or even impossible in the actual.

But this is not the same virtual as the one mentioned a couple of paragraphs above, associated with Deleuze. This digital virtual is a crude parody of Deleuze's virtual: it replaces the open-ended potential that constitutes that philosophical virtuality with a set of discrete and determinate possibilities, and in place of the as-yet-unmade pathways of a world becoming something else it substitutes the knowable selections of a choice, a choice that always boils down to 0 or 1.[1] The digital can institute an ontology of its own, can tame the wild indeterminacy of the actual, because it creates a space apart, a microcosm modeled on a post-Enlightenment fantasy of perfect reason, where everything is knowable and everything makes sense.

Ontology is not a solved problem. Though scientific discourse may presume, in some contexts, to offer a bottom line about the way things *really* are, there are various and even competing scientific descriptions of the real, plus plenty of other expressed or tacit metaphysical theories (reality as god, or spirit, or dream, etc.), and the community of philosophers achieves no greater consensus than do laypersons on this matter. Some beliefs about the nature of reality are fairly consistent with the ideology of the digital, including the popular idea of material monism held up as the secular default metaphysics of our age. An ontology that apprehends the actual as made of posits, which was arguably the core claim of the early- to mid-twentieth-century philosophy of *logical positivism*, would likely provoke little tension between the actual and the digital, however it understands their differences. If the actual exhibits a positivism, then we would expect the digital to do a pretty good job of mirroring or capturing it.

But a nonpositivist outlook, or even the inclusion of some nonpositivist aspects within a given metaphysical position, presents a significant challenge to a digital perspective. Though Western philosophy has, throughout its two and a half millennia, been repeatedly drawn to forms of positivism, beguiled by its satisfying clarity and confident assurance of a bedrock certainty, there is sometimes also, within those very positivist philosophies, an embrace of antipositivist or nonpositivist dimensions, a recognition that the assertion of phallic presence does not adequately describe everything in our experience. For example, if the monotheist god represents the ultimate posit, "I am that I am," that same formulation betrays a metaphysical obscurity that would seem to defy positivist capture. Antipositivist ideas, philosophies of *difference,* dot Western intellectual history from the Classical era to the current moment. In keeping with their essential priority on difference, philosophies of difference differ amongst even themselves and do not adhere to some common ideological commitment. If they tend to be antipositivist, they manifest this antipositivism in a variety of ways. The following paragraphs that describe in overview a few such antipositive philosophies illustrate but do not defend that variety of nonpositivist thought. These elliptical accounts do not pretend to a comprehensive or critically examined explanation, but instead aim to demonstrate the recurrent appeal of a worldview that prioritizes difference over the positivism of identity.

The pre-Socratic philosopher Heraclitus, of the fifth century CE, offers a suitable ontological starting point. His philosophy of becoming, surviving only in fragments of lengthier works, articulates an ontology that emphasizes flux as the basic way of being of all things. His most celebrated contribution is the insistence that "no man ever steps in the same river twice." A positivist might understand this popular epigram as an observation about an accidental property of the river; that is, we might read this sentence as suggesting that there is a self-same river ($\alpha = \alpha$), but that the river always has different water coursing through it such that it is never quite the same as it was a moment ago. But to appreciate Heraclitus's aphorism as a metonymic claim about the world in general (and not just about flowing water) and as a statement of fundamental ontology would motivate a more radical reading: there can be no such thing as the *same* river, for everything is change, constitutively change. If the world is made of change, if all that is is change, then this leaves no room for sameness, for permanence, except the sameness and permanence of change. The image of the river serves as a bridge to a more basic and foundational idea: everything is like the river, essentially constituted by flux.

Where there is only change, where the only permanence is flux, there can be no unassailable rule, no transcendent fixity. Every principle is negotiable, every pattern modifiable, every repetition a potential deviation. This is the basic condition of *contingency,* wherein the only rule is that there are no rules. Quentin Meillassoux, a philosopher of the twenty-first century, states it most bluntly: the only necessity is the necessity of contingency. (Fifty years earlier, Wallace Stevens had anticipated this radical philosophy in a famous poem: "The only emperor is the emperor of ice-cream.") The question at hand is: How does the real measure up to these accounts of it?

Quantum Computing

Consider the case of quantum computing, no less loyal than is conventional computing to the rules by which it operates, but governed by the more unruly mechanics of quantum physics. Its rules do not wholly determine each individual case, but operate strictly only at the statistical level over many cases.[2] Even so, this is no radical contingency, for the individual cases resolve to one or another value, a multiplicity of possibilities but no florid potential for the unanticipated. Under those circumstances, the quantum computer does not reliably yield a correct result of its calculation each time, but generates an outcome with a certain probability of being correct. Each calculation, each algorithm, must be carried out multiple times to provide a sufficient confidence that the recurrent result is not a statistically unlikely fluke. Moreover, this is why quantum algorithms typically conclude with an operation called a Discrete Fourier Transform

(DFT, though actually it is a quantum version of the DFT, the QFT): a mathematical transformation that takes a sequence of entangled qubits that represent a statistical distribution of possibilities and puts them into a related state that identifies regular patterns within that distribution of possibilities. Calculations must be performed numerous times to achieve a relative certainty in the final result, and even that relative certainty does not match the confidence of the result of a conventional computing machine.

Just as classical computing institutes a rationalist, rule-bound regime, so quantum computing imposes rules in order to reign in the more wild quantum effects. The coherence of a quantum system requires that it be isolated, or it cannot serve computing purposes. As quantum phenomena cannot be completely segregated from the surrounding environment, there is always a risk of decoherence during a calculation process. Though information can never be lost in the quantum realm, a quantum computing system decoheres by "discarding" some of its information into the surrounding environment, and the original qubits, having given away some of their information, no longer contain the "correct" result of the performed calculation. The larger system—qubits plus environment—still has all of the information, but the computation, which relies exclusively on the qubits, is no longer reliable. Quantum computing functions by imposing rules to exclude the contingency of an ever-expanding system of information conservation. It is therefore not a promising rejoinder to the obsessive rationalism of classical computation.

Friedrich Nietzsche is a point of reference for nonpositivist thought of the twentieth and twenty-first centuries. His notion of the "eternal return" (for example) is often mistaken (even by Zarathustra's animal friends) to indicate the elimination of contingency through a perfect (necessary) conformity to a past model: the eternal return as an endless cyclical (hence necessary) recurrence of the same. But Nietzsche mocks this reading, instead welcoming the eternal return as a maximum of contingency. For what returns, in an echo of Heraclitus, is always again the contingent, always and everywhere the freedom of the new, the un-dependent, creativity as an ineluctable principle. Contingency gives the world texture, sets things in motion, else the world would effectively grind to a halt in which nothing new, nothing meaningful happens. Nietzsche recognizes that contingency honors no rule but its own, so that every rule is threatened by violation and displacement, acts of iconoclasm and revolution that clear the way for creativity, and for Nietzsche, progress toward the salvation of humankind. In the next few paragraphs, witness four post-Nietzschean theorists, each offering an ontology that retains this crucial place for contingency.

To insist that there is only contingency is tantamount to the claim that nothing is necessary, that everything depends on other things, which also depend on other things, and so on, the whole universe depicted as a ramifying network of conditioned sensitivity. This is indeed the metaphysical picture drawn by many thinkers and codified, as mentioned above, by Meillassoux. Meillassoux's scholastic argument is dense and complicated. At its crux, the argument demonstrates that to oppose the claim of contingency requires in effect that one implicitly rely on that very claim, and is therefore not a consistent position. Meillassoux also acknowledges that his argument must confront a seemingly intractable problem: to think an absolute contingency is to think what cannot be thought, for whatever is in thought is conditioned by the limits of thought. But if his path to his conclusion is tortuous, and if he leaves some pressing problems unresolved, Meillassoux nevertheless succeeds at constructing a strong proof of his thesis, "first, that contingency is necessary, and hence eternal; second, that contingency alone is necessary" (65).

He does not shy away from the implications of this metaphysical shocker. If there is no necessity (except the necessity of contingency), then nothing is certain, everything becomes unmoored, undermined, and we must reckon with the repeal of all of the everyday, unstated assumptions about the future that support our every action and thought. We confront, as Meillassoux puts it, *chaos*. The relevant cliché is the sunrise: if the sun rising is a contingent event, if it might or might not happen with no necessity supporting either alternative, then this clarifies the unmeasurable stakes of Meillassoux's discovery even as it seems to put his conclusion beyond the reach of our thought. That is, to recognize the radical implications of the sole necessity of contingency is also to recognize that one cannot fully absorb those implications, for thought itself and any possible action all rely on the reasonable expectation that things will be more or less the same as they have been, whereas contingency reveals that no such assumption is well founded.[3] Meillassoux's is evidently a world without god, a world without transcendence, a world without a firm place to stand.

Whereas Meillassoux plays up the atheism of his argument, emphasizing the sense of a universe radically unhinged or disordered by the repeal of every rule, other twentieth-century theorists discover not so much the threat of chaos but the source of freedom, a vital or creative foment, set loose by the abrogation of necessity. When no rule is absolute, every event harbors the freedom to choose a new path, to venture into the unexpected, the unconditioned. This creative freedom, the freedom to supersede any rule by invoking a higher one or to cast off rules altogether, this freedom at the heart of ontology, becomes for Deleuze the affirmative principle of all becoming, for Jacques Derrida the very possibility of making meaning, and for Heidegger the safeguard of self-determination.

Heidegger's treatment of contingency may be the trickiest and most ambiguous, in part because it unfolds in the context of an ontology in which the being of the world is tied inextricably to the being of *Dasein,* the one who is in that world, experiencing that world. In fact, a chief motivation for Meillassoux's argument about thinking beyond the limits of thought is the sense that Heidegger's phenomenological ontology is too limiting, that Heidegger corrals his philosophy within the perspective of an experiencing person, *Dasein,* and so forecloses any consideration of a world unconditioned by the philosopher's thought of it. That said, Heidegger's phenomenology does not rely on a simple idealism wherein the world would be an artifact of the subject's thought, for a chief characteristic of *Dasein*'s world is that some of it is unknown to *Dasein,* outside of *Dasein*'s thought. Part of *Dasein*'s experience is the very limit of that experience. Complementarily, Heidegger also wards off the threat of solipsism, as *Dasein*'s world essentially includes the social: being with others in a shared world is a fundamental and indispensable part of *Dasein* and its world.

In Heidegger's analysis, being with others, though it safeguards the fundamental sociability of *Dasein*, also offers to *Dasein* the relative ease of the mundane or routine. Faced with the existential questions that constitute human being—What should I do? How should I act? Who am I?—*Dasein* generally falls back on available conventions in its relation to itself, to others, and to the world around it, conventions always at hand due to *Dasein*'s essentially social nature. Anxious about its place in the world, *Dasein* retreats into the placid comfort of the crowd perspective, doing "what one does" and saying "what one says." This diagnosis of *Dasein* turning away from its ownmost individuality to adopt instead the generic perspective of what one does is perhaps the foremost ethico-ontological problem presented in Heidegger's great work, *Being and Time*. And his response to this challenge, subject to much interpretive debate among Heidegger scholars, is that *Dasein* must embrace its *unique and unconditioned possibility,* which it encounters only when facing the prospect of its own nonbeing; that is, *Dasein* summons contingency, its way forward that can never be given by a conventional rule, by staring into the ultimate certainty of its own death. Death, as it looms before each of us, awakens contingency, frees *Dasein* from the warm cocoon of tranquil everydayness, so that *Dasein* must decide and embrace its own-most future, must choose to be its authentic self, having loosed the bonds of normalcy that relieve the pressure of decision. In Heidegger, contingency frees *Dasein* to take responsibility for itself, which is both a terrible burden and the only way to be who one most of all is.

Much of the debate around *Being and Time* revolves around how to understand the contrast between inauthenticity, in which *Dasein* relaxes within the normalcy of behaving as one does, and authenticity, in which one anxiously confronts the contingency of one's own-most possibilities of being. Do these modes of being constitute a choice between two different ways to behave, or are they both essential aspects of *Dasein*'s being, somehow simultaneous in spite of their opposition? Obviating this interpretive (and ethical) dilemma, Deleuze offers contingency an even more prominent position in his ontology, in part by rejecting Heidegger's phenomenological emphasis on *Dasein*, who is in effect a human subject, and so paints Heidegger's worldview with the brush of humanism. (Recall the humanism of Heidegger's critique of technology, which complains not that technology generates untoward effects, but that the technological point of view shuts out other relationships to the world, and so severs us from our own humanity.) Instead of an ontology deriving from *Dasein*'s special human perspective, Deleuze acknowledges a multiplicity of perspectives, sometimes rooted in persons, but not necessarily tied to humans or even to living creatures. Searching for a term that could capture the complexity and dynamism of a perspective but that also indicates a breadth well beyond the human, Deleuze often talks about a world made of or made by *machines.* He advances a universal vitalism that discovers little *machines* everywhere, functioning for Deleuze as protosubjects. This vitalism invests every synthesis with its own way of being, such that, when disparate elements in relation achieve a stability, there one finds a perspective, a protosubject no longer connected to persons or even to the domain of biological life. The universe constantly generates new perspectives, new ways of being, and this incessant creation, often rendered invisible by a positivism that recognizes only sameness, itself results from an originary contingency. The universe is reborn at every instant out of difference, the unbounded, the unruly, which is to say that, for Deleuze, contingency is the creative principle behind all things and all change. The core principle of the machine that manufactures the world is contingency.

One might imagine that, casting the world as a rich field of creative machinery, Deleuze would embrace the digital as a demonstrably fertile sort of machine, one that, through artificial intelligence, deep learning, and other advanced applications, begins to manifest explicitly those protosubjectivities that he discovers in all matter. But Deleuze refuses this conflation of *machine* with *mechanism.* His machines are heterogeneous syntheses, self-organizing collections of forces, materials, and ideas that form temporary stabilities and that constantly remake themselves. Persons and animals are machines, but so are stars and galaxies, as well as economies, academic disciplines, commemorative events, and science fiction novels. There is no comparison between these semistable part-objects, these fields of potential that reshape the world around them, and the self-contained, well-defined, rule-based, step-by-step advance of a deterministic digital algorithm, which includes no accident, no contingency, but only the encoded intentions of its designers, driving it toward its preconceived moment of termination.

Derrida, too, structures his writing as the preservation of a spontaneity that rules out preconception, the denial of origin and the refusal of telos, such that he holds things open, much as do Deleuze's machines. Derrida's preoccupation with language redirects the intensity of his ontological claims, for he frequently problematizes the distinction between how things are and how we talk about how things are. That very equivocation between discourse and world marks a space of contingency: language for Derrida is fundamentally *playful*, meaning that it enjoys a creative leeway, a supplement that exceeds any rule that would circumscribe language once and for all. Of course there are rules, there are also always rules, but no rule is sacred, and any might be in play. Anchored to the world in countless ways, language also escapes those anchors and slides across the world's surface, forming something new and unanticipated: meaning. (*Making* sense is no mere figure of speech, for sense is created in the meeting of language and world.) Derrida affirms in the same breath the creative generation of meaning and the creative constitution of the real, and so locates contingency, the unrestricted freedom of play, at the core of both meaning and being.

These post-Nietzschean promotions of contingency declare its essential role in—even equivalence to—freedom, which can be both terrifying (in Meillassoux's extrahuman ontology or Heidegger's weighty assignment of responsibility) and liberating (in Deleuzian affirmation or Derridean play). Contingency denies necessity, freeing things and events from the rule of rules. It means that no rule is transcendent, that any rule might be refused, altered, ignored; as a point of inflection, contingency has the power to invent or invoke a superordinate rule. One might worry that such an unruliness would not make sense, that there can be no sense in the face of contingency, for contingent things and relations cannot promise the stability that would afford sense a foothold. If cause and effect are imperiled by the intrusion of contingency, if relations in general are rendered accidental, expressing mere happenstance rather than underlying order, then there can be no sense, for sense is the articulation of an order, a reason beyond the particular fact of a given circumstance.

But this is to confuse contingency with chaos. Though contingency may be at odds with a cosmic order (the monotheist god or an ultimate, inviolable rule), it does not destroy all rule or regularity. After all, we can and do make sense. We recognize patterns, we credit explanations, we trust stable relations, we plan and carry out action using inductive reasoning, and we believe in causes and their effects. How do we understand sense-making while still preserving an essential and ubiquitous contingency? Or, flipping the question around, what kind of sense is available in a world with haloes of indeterminacy, precarious relationality, and unstable identity, a world of contingency? It is a sense made, a sense forged, a sense won, rather than a sense found or reported. To make sense is to establish order in defiance of its nonnecessity. Only because nothing need make sense can anything make sense.

To make meaning is to sew together language and world according to some order. Order does not await its discovery; if order were already there, in the world, in the state of affairs, then it would need no making. A fixed object or event, a link in a chain of necessity, does not and cannot make sense. Even Claude Shannon's decisive mathematization of communication insists that information depends on entropy, that the more order in a given situation, the less information there. Sense or reason can be made only where there is already a question, where contingency erodes order to leave something in play, something undetermined. The sense of a world conditioned by the necessity of contingency does not derive from respect for a rule. It is not the sense of categorical belonging grounded in identity. Under the singular rule of contingency, it would not make sense to locate each thing or event as a species of a general type, for this is precisely to hold steady, as against the admission of contingency, the type, as well as the criteria of determination for belonging to that type. Nor would sense in a contingent world be a matter of conformity to pregiven rules. If the rules are understood as fixed, then it would not *make* sense to cite them, but would only ratify their inexorability without offering any additional insight. Rather, a sense that respects the spontaneous production of reason and accords with the play of contingency would select a reason anew at each junction, each call to make sense.

This is a sense renewed at every instant, a sense with lability and dynamism to match the foment of a world set free, a sense produced in concert with the lively world it explains. Could such a sense be sensible at all? Can rationality, the capacity to make sense, flex its brittle chitin to leap and dance, to produce the sense of an unruly world? But white-bearded reason, supposedly so staid and unassailable, already answers to the manic tarantella of the sense of contingency. What else does Kant, that most sober of philosophers, mean when he insists that reason is absolutely spontaneous?

The spontaneity of reason claims for reason an overarching authority, that reason shall be its own arbiter, reason as the final judge of its legitimacy. Sense, in that sense, stands above any rule that would constrain or delimit it. But that does not mean that reason is arbitrary, that anything at all would be reasonable. Reason that can refuse any particular rule sounds simply unreasonable, a partnership with chaos, no sense but the meaningless. If sense depends on the precarity wrought by contingency, if reason is produced along with the order that invests it, then its judgment is no rubber stamp. Even if sense determines its own validity, it holds itself to an exacting standard, a spontaneous criterion that nevertheless imposes real constraint. Reason might appeal beyond any existing rule, but it must still honor its boundaries even as it reserves the right to redraw them. How can we understand those boundaries, those constraints on reason, if they lack the power, the steadfast insistence to keep reason in check? What does it mean to have a rule if that rule can be broken?

To resolve these antimonies, we must return again to the meaning of contingency. Contingency frees the world from the death grip of absolute rule, but if it borrows from chaos its capacity to overleap any rule, it does not altogether abandon reason, but sustains the conditions under which sense can be made. Contingency unhinges things and events from the world, threatens to sever any bond, but it does not overturn order *tout court*. Instead, contingency allows things and events their contexts, and it leaves intact most of the ties that it also threatens, holding in abeyance its most disruptive potential, which is always real but not yet actualized. Another name for this precarious organization of sense is *overdetermination.*

A Meshy Reality

Positivism depicts a world made of autonomous individuals: each thing is first of all and originally itself, and things bear relations to other things only secondarily or after the fact. Sense in such a world would be a matter of connecting the dots, selecting the thing or things responsible for some event or occasion. To disavow positivism, to substitute an alternative image of the real, is to elevate relation to a primary position, to start with a relatedness that precedes and gives rise to the things it brings into relation. These are not fixed individual points, but textured fields of differentiated relation: an antipositivism would call for a wholly different topology. Instead of individuated posits with binary relations as the ontological bottom line, there is a differential mesh of the finest threads, condensed here and there into knots and whorls that demonstrate a partial stability as a complicated collection of relations.

To make sense, one must select some of these threads, trace them across this variegated mesh, draw them from the field that surrounds each thing with potential, potential directions, potential connections, potential becomings. The mesh is the haze of relations, the many ways that things encounter each other or communicate across their distances: association, causation, resonance, appearance, family, gender, quality, temporality, intuition, proximity, reaction, rejection, and so on. Relations in the mesh are not so much between two things as over different areas, from knot to knot, or across a fold, or within a perimeter. Enmeshed things don't relate as posits that preexist their relations, but rather they come to be, achieve a degree of density or contortion, coalescing already amidst relations. Relations determine their objects even as they develop in themselves, a web that does not capture and paralyze the world, but realizes the creative freedom to go where it will.

Positivism, jealous of the creative power of a living mesh, builds its own model, "the network," beginning, as it must, with individuals, which are called "nodes" in the context of the network. Networking opens all sorts of new complexities of positivist mechanism, delocalizing to the point where place itself becomes diffuse (for the network is everywhere at once), but networks, for all their variable topology, remain structured by biunivocal relations among nodes. Whatever degree of complexity a network might demonstrate, it is different in kind, not the same variety as a mesh of difference. In a network, the nodes and relations are self-identical and fall under species within a genus, whereas in a mesh, the threads are heterogeneous, not so much elements as events, acts of absorption and cleavage, invention and extension, more threadings than threads. Never linear, never stable, each thread is itself a dynamic meshwork of finer threads that exhibit all the same kinds of variation—dead ends, conjunctions, multiplications, discontinuities—so that no thread can be identified as a bottom line, a base element, or rather a thread might be gathered into a bundle to constitute a thicker thread, but this relationship of element to structure is itself relative, and the thinner thread is also a structure with other threads as its elements. Complexity, in this antipositivism, is the only rule. That is what the network cannot simulate, however numerous, however plentiful its nodes and relations. A network, even at its densest and most dynamic points, still prioritizes the individual nodes and their (secondary) relations, but no number of individuals ever achieves the complex heterogeneity of the mesh. (This is, in a nutshell, this book's critique of the digital.)

Sound Art

David Dunn is an environmental sound artist and acoustic ecologist who takes the somewhat unusual stance of choosing to interact with the environment that also serves as his recorded object. His art-based research underpins his attempt "to recontextualize the perception of sound as it pertains to a necessary epistemological shift in the human relationship to our physical environment" (3). Whereas vision supports a positivist outlook by localizing and defining its objects, the medium of sound, which is diffuse and nonlocal, offers Dunn an antipositivist relationship to a highly interconnected world. "In Buddhism," he writes, "the concept of *Sunya* (a Sanskrit word translated as 'emptiness') describes the complex chain of connection that forms the world. Each 'thing' is so densely connected to everything else that it resides nowhere. We cannot isolate the thing from all the states of matter or energy that preceded it or to which it will become" (4).

Dunn affirms contingency, the condition in which everything touches everything else, and recognizes that strict identity becomes unsustainable under that condition. Always and already in contact with the birds, bugs, plants, people, instruments, and other phenomena in his audio recordings, it would make no sense to Dunn to claim a neutrality, as though he too (and his gear) did not also vibrate, contributing to the sound in space at that moment and engaging the meshy fibers that connect him and his recorded object to the entire environment. Any recording includes all of the circumstances of its production, everything that touches the scene. He emphasizes that sound exposes the imbrication of each thing in every other, whereas vision separates things into distinct entities with relations only after the fact: "When we look at the world, our sense of vision emphasizes the distinct boundaries between phenomena. The forward focus of vision concentrates on the edges of things or on the details of color as they help us to define separate contours in space. . . . The sounds that things make are often not so distinct and, in fact, the experience of listening is often one of perceiving the inseparability of phenomena. Think about the sound of ocean surf or the sound of wind in trees. While we often *see* something as distinct in its environment, we *hear* how it relates to other things" (1).

The digital captures sound only by stilling its diffusive and restless traversal, giving it strict definition and clear boundaries. In digital contexts, even environmental sounds, such as birds chirping or a garbage truck backing up, are constructed effects, part of a world of human intention. Instead of emerging from *Sunya*, the chain of connection that forms the world, a digital sound is generated as part of some object; it is a behavior of something, whether an interface object or button, or just the object that is the digital general surround, the operating system or the sonic background of the simulated world. In the digital, nothing arises organically and everything must be deliberately inserted by algorithmic intervention.

Fruit makes a fine example, helping to reveal a world that admits both the destabilization of contingency and the multiscalar, dynamic order that subtends reason. All the parts of a lemon, for example, respond to the same interwoven set of forces, such that each part—the pulp, the peel, the seeds, the shape, the color, the flavor, the chemical constituents, its economic situation, its role in cooking, its cultural resonances, its aesthetic possibilities—all of these aspects of the lemon are tied together by its histories, species history, natural history, individual history, its reason that makes it what it is, determining in a coordinated evolution its form and content, its appearance and many behaviors. To answer the question of why a lemon is sour, to make sense of its sourness, is to choose one or more threads from within that contingent but coordinated history, that *overdetermined* set of possible reasons, factors that provide the contextually appropriate response. It's sour because it developed in a certain climate, determined by geography, with certain available nutrients, plant forms, relations with fauna, and so on. Or it's sour because it has a relatively high concentration of an acid with a molecular structure that encounters responsive structures on part of the tongue. Or maybe it's sour because, well, lemons are sour and it's a lemon, or because it's finally ripe, or because certain citrus fruits tend to be sour, or because you just ate something salty. These multiple reasons, a list that can continue to expand indefinitely, manifest the overdetermination of reason, which is how contingency intervenes in this case. No rule decides in advance which reason will be relevant in a given context, and one never knows where a thread will lead, what it might pass through as it marks the relations that determine the lemon and its sour taste.

One could as easily cite examples from the human world as from the natural. In the actual, the real world, things and their parts, their forms and their contents, are not independent, for they answer to the same reason, a shared history with the form not of a simple line or even many lines, but a tortuous mesh whose only rule is the antirule of contingency. The roof, walls, and floor of a building respond to a complex logic of habitation, shelter, property, gravity, materiality, human scale, aesthetics, human needs and patterns of motion, sleeping, and living. It's not a reductive or simple logic,

and it's not wholly deterministic, for it leaves much freedom of relation among various dimensions, but it guides all the parts of a
building, ensuring their coordination and their interdependence.
This mesh logic, expressive of its contingent condition, assembles
the world, and so sets forth the relations from which sense can
be constructed. The world does not await its sense, for world and
sense are enmeshed each in the other, arising in the same becoming, the same knotty weave of threads.

Generated and regenerated by contingency, the mesh exhibits both complex stabilities and a ubiquitous precarity. Any resonance on the mesh confronts a contingency that might tear it apart
or remake its relations or incorporate it into still larger forces and
traversals, but each resonance, fleeting or standing, also ties together so many threads in a complex tangle, such that each thread
enters into numerous relations with many others and the bundle
holds together only by virtue of all of its contributing lines. This
is how reason can be at once overdetermined and unruly, a ten-
segrity of potential factors locked in a tight embrace. The characteristics of things and events in the actual interrelate across many
dependencies, giving those things and events a kind of inertial
weight but also opening them to dramatic change in response to
small motions, the reorientation of the intense forces locked into
those relations. Caught in their relations, things in the actual cannot be easily manipulated independently, resisting a combinatoric
instrumentalization.

This account of ontology exposes both its frenetic dynamism
and its stolidity. Things and events come to be as knots of relation
that achieve a temporary (though sometimes enduring) stability;
the world is born and always reborn out of a fundamental generativity, the eternal return of renewal. But, arising as a stable structure
caught in the mesh around it, a thing's parts or an event's moments
are all tied together by that meshy complexity of interrelation. The
lemon and the building are not merely aggregated parts of independent qualities and dimensions that have been pieced together
according to some menu of ontic production. Rather, the parts and
characteristics of the lemon, of the building, of a concert, of a political revolution, of a life, of a species, of a molecule, of an institution,

are all parts and characteristics that respond to each other, such that piecewise change is unlikely or even impossible. Careful genetic manipulation might enable us to grow purple lemons, but that fruit would likely taste different from the original yellow lemon, its rind a different thickness and texture, its ideal climate and soil to be found in somewhat different geographies. Notwithstanding the churn of ontogenesis, that *teeming tangle* of disindividuated threads (more threading than thread), things and events retain a kind of integrity, a complex and porous internality that ensures that any sense is overdetermined, tied to too many factors, swaddled or punctured or pulled along by too many threads.[4]

Where can we locate contingency in this abstract description of the universal ontology of the actual? From one perspective, contingency is the (sole) rule of generation, the creative moment that sets the threads in their wild growth, entwining and articulating them, to invent, and always reinvent, our meshy reality. Contingency twists things together and splays them apart, establishes rules and then breaks them in favor of new ones; it grows and distends the mesh, guaranteeing its irreducible complexity and its incessant renewal. Contingency underlies ontology broadly, giving rise even to the whorls and creases, the knotty masses in tight and steady resonance, producing in effect provisional centers, more or less stable identities. This partial or precarious stability, more or less stable identity, does not yet determine the perfectly discrete individual of positivism, though it is a signpost on the way there. Stabilized relations within a knot of threads dampen the wildest fluctuations of the mesh around them, conserving their stability and intensifying the resonances among those relations. Contingency is not absent there, but it is muted or ameliorated, its generative force evident mostly around the edges, in the margins of insignificance too fine or too irregular to stabilize consistently. It would take an act of idealization, the imposition of an absolute cleavage absent from the world but available to a subject, to turn these proto-individuals, these knots of semistable relation, into self-identical posits, cleansed of all contingency. As the next chapter describes, that act of idealization is the distinctive gesture of the digital.

Making Art

An event that honors all rules would be expectable if not always expected, and so could not be wholly creative. But consider the paradigmatic creative activity, art-making. An artist always generates her artwork through accidents. No matter how technically proficient the artist, the artwork happens in negotiation between some intentional gesture by the artist and a medium that necessarily, to some degree, resists that intention or redirects an applied force, or simply defies expectation. Which is to say that all art proceeds to some degree as improvisation.

A dancer may have developed astounding precision with his body, including an ability to disarticulate his parts and use them seemingly independently. A calligrapher may employ remarkably refined brush or stylus control such that we (famously) regard the point of application of paint (or ink) to the writing surface as though it were part of the calligrapher's body, even an unusually responsive part. An oboist might practice scraping her reed to achieve a fantastic consistency, with an embouchure to match, uniting her and the instrument under a single will. Yet, in each case, the aura of the work, whether performed or installed, whether witnessed live or reproduced mediatically, depends on a thin edge of resistance, where the tool or the context refuses to cooperate, demanding spontaneous adjustment from the artist. The dancer finds his left leg a little more achy than at yesterday's rehearsal, or the newly swept stage floor a bit smoother, and this discrepancy calls for compensatory (though possibly unconscious) responses, which may alter, however slightly, the entire performance. The calligrapher always engages a writing surface that absorbs ink or diffuses paint with an incalculable variability, necessitating constant adjustments of pressure, angle, speed, and so on, and producing a result that at its edges exceeds her intention, a sliver of contingency. The oboist may, without thinking of it, alter her playing in response to the unusually humid concert hall, but that humidity will vary even as the concert proceeds, and a master musician enjoys that tiny variation as an opportunity to invent, enriching the performance in unexpected ways, as her instrument surprises her. In each case, contingency denies an absolute control, enforcing through its accidents an improvisatory margin that inflects even the most refined and exact gestures, an improvisation that marks a genuine creativity, an artwork forged by artist and world in a salutary and ongoing cooperative practice.

Under these conditions, in which every question and every reason refers to a complex tangle of ever-expanding relationality, the production of sense (or the provision of reason) can be only a matter of selection. The world is overdetermined, governed by many rules in relation to each other, such that reason requires choosing the relevant or appropriate cause, tugging on the most apposite bundle of threads. Contingency takes on a new emphasis in this figure: though it still functions as the impetus to the growth, diminution, conjunction, and partition of threads, it should also be understood as the always available potential to connect a thread that had not previously been in consideration, to invoke a new structure, a new rule, that not so much negates the old ones as entwines and redirects them. Not the negation of all rule, but the refusal of any transcendent rule, that is the meaning of contingency in this meshy world, and thus it replaces the principle of sufficient reason with a new principle fit for a world of overdetermination, a *principle of abundant reason.*

The principle of abundant reason means that one cannot decide in advance which reason will prove salient, which explanation will prove most sensible or compelling. The question is to choose the explanation (or explanations) that are most apt in a given context. Sometimes it makes the most sense to describe the start of a war as the result of long-simmering distrust brought to a head, sometimes as the immediate reaction to an act of aggression, sometimes as the outcome of a fundamental incompatibility between two national ideologies, sometimes as an economic opportunity seized by one side. And one might prefer in some cases an explanation that appeals to the warlike nature of humanity under stress, or a political theory of cyclical tension and release, or even an unusually intense period of sunspot activity. All of these explanations might simultaneously be correct, but depending on the question, only one might be most relevant or informative. There is no rule in these cases, or no transcendent rule, because there are many rules; and how they will reshuffle and recombine, and what further rule may be invoked to account for those shufflings and combinations—these are contingent matters, the entry of contingency.

A world of abundant reason accounts more fully for the distinction, marked above, between the freedom of contingency and the chaos of an irrational universe. The refusal of any transcendent rule might seem to undermine the very possibility of order, to tilt toward the arbitrary or chaotic. But an abundance of reason ensures both the possibility of a partially or locally stable order and the ontological latitude to invent a new rule or invoke an unfamiliar one. (Thus is the abundance of reason also an invitation to the spontaneity of reason.) Caught in numerous and knotty threads, a thing cannot simply alter one or another of its ways of being, for they are all interconnected. Combinatorial analyses, which treat things and events as aggregations of ultimately independent and possibly interchangeable parts, cannot account for the twining of the many threads that come together not in combination, for they are not individuated, but in a fractal geometry of the mesh. The stability of things derives not from a central or core identity that polices the margins and maintains definition, but from resonances among all of a thing's many facets that hold it together, relating each of its aspects to the others. Contingency produces local and momentary stability, but also ensures that its stability is under threat, never guaranteed, given as a becoming, which may be sudden or glacial, vast or localized, consequential or insignificant, momentary, recurrent, or enduring.

A thing is the site of a détente, an uneasy ceasefire between the boundless potential of becoming and an inertial resistance to change. The same mesh that thickens to constitute a thing and hold it together also always carries the potential of a radical reconfiguration, a new thread or a rethreading, that reverses figure and ground, or breaks a whole into parts, or otherwise effects a change of status that could be sudden or imperceptibly slow, could alter the course of one's day or the course of the whole universe. The mesh ties a thing into a complex knot that maintains a consistency and coherence, but also mandates that, when the thing changes, it usually changes in many ways, for it is a highly interconnected bundle of forces, not a combinatorics of independent parts.

Cars That Drive Themselves

The ethics of self-driving cars is a current issue in digital culture. A software-controlled car may at some point have to make a *decision* about what to do in a case of only bad choices, like an actual instance of the typically hypothetical "trolley problem." Steer one way and hit a man carrying a baby; steer the other way and hit a crowd of teenagers. There is no good choice, and much ink has been spilled over the question of how to program the car to make the best choice, the difficulty of which is compounded because we humans don't always agree on what the best choice would be.

The car has one significant advantage over a human being in this situation, in that it can make split-second decisions that can, notwithstanding the very small time frame, include sophisticated calculations of the risks and potential payoffs of the available choices. If we could identify reliable and unambiguous rules of moral choice, we could program those rules into the car's algorithmic control center, efficient enough to calculate the correct outcome even in the few milliseconds available in emergent situations. Humans, on the other hand, enjoy a different advantage, the possibility of incorporating new information, taking account of unforeseen options, and even discovering a new frame for viewing the problem. Indeed, the potentiality of a new frame is arguably the very condition of ethical decision making. That is, the possibility of reprioritizing any given rule or of inventing a new rule is a human capacity unmatched (and unmatchable) by any digital machine. For, once programmed, the automobile software is bound by its programming, able to take into account only those factors already anticipated by the programmers, and never able to reframe the problem, treating it according to criteria unanticipated in the software.

This is fundamentally an illustration of the spontaneity of reason. People can always discover or invent a new perspective, one that upends older perspectives, for no rule is inviolable. Ethics, as a kind of reasoning guided, like all reason, by intuition, is spontaneous. (This is why it is always possible to convince oneself of the justice of any choice: it is part of the practice of ethical reasoning to invoke new frames of reference, to appeal to a higher principle, even in bad faith: "I am normally a man of my word, but in this instance, breaking my promise is clearly the right thing to do.")

We could call it the "Law of Digital Hermetics": the digital is always bounded. There is always some frame, some rule, some fixed boundary that the digital cannot break. Put otherwise, the digital always has a definite inside, delimited or defined by this inviolable boundary. The digital constructs its own world, apart from, if thinly connected to, the actual, and the borders of that world are enforced absolutely. A person can alter that most exterior digital boundary, add in new code or alter the existing code to supersede the nth layer of rules and impose an $n+1$th layer that overwrites the existing bottom line, but the hermetic internality of the digital always returns, now in the form of an even more remote frame.

How stands an abundant and spontaneous reason as regards the digital ideology? An ontology grounded in contingency counters digital ideology at each point. Contingency rebuts positivism by virtue of its principle of becoming; the world is not determinate, does not rest in static being or ratify an assertion of independent individuality (i.e., a posit), but instead always must become what it is, everywhere bound into and propelled by a mesh of relation. Nothing in the mesh of the real remains still and isolable, for everything is run through with relations that both sustain and evolve what things there are.

This poses an impossible hurdle also for rationalism. Rationalism depends on a stable and parsimonious order, a totalizing harmony, whereas contingency threatens always to institute a new order, wrapping things and events in a haze of relations that include disharmony and tension and that feather into insignificance at the margins. The principle of abundant reason might seem numerically to guarantee a sufficient reason, but in fact they are opposed: the principle of abundant reason invents reasons anew or draws into play a rule that reorders the old ones, making a joke of a sufficient reason that would trace the stable order of all things. Rationalism withers, as the world does not have any final or ultimate sense. Rather, sense must be made tenuously, contentiously, and only a bit at a time, for sense is always contested and eventually relinquished. No order of causes and effects, no master plan can emerge from the chaosmos (chaos + cosmos) of the world, and so no rationalism can grow beyond the most local and temporary examples.

Fractal geometers talk about measuring coastlines, a ready illustration of the "fractal geometry of nature," to quote the title of Benoit Mandelbrot's beautiful but sloppy guide to the subject. One might measure with a one-meter rod, but a one-centimeter rod will yield a longer coastline, and a one-millimeter rod a still higher measurement, because the smaller measures take account of more variability, more jaggedness in the line dividing land and water. The roughness of the coastline, which has a fractal dimension that indicates more or less how jagged it is at each scale of measurement, follows no fixed rule or set of rules. A computer can simulate that roughness by instituting a rule in the form of an algorithm, and can even reproduce the appropriate fractal dimension (up to its limit of resolution or memory) with a properly designed rule, but what it will always fail to replicate, to simulate, is the way in which the coastline's shape exceeds to some degree any rules, for it is absolutely contingent, an artifact of an infinity of forces interoperating over ages of erosion and accretion.[5]

The form of the unruly coastline differs crucially from randomness, which would admit no reason. On the contrary, this unruliness complements and invites sense or reason: unlike a random arrangement, the coastline's shape has reasons but no rule. Each part of the coast is shaped the way it is due to that congregation of forces that have shaped it, giving it a reason, but that reason is local and contingent, adhering to no higher or overarching rule. Contingency thus underlies sense but does so from the margins, not by a direct interjection into our sense-making, but by a contextualization that makes sense possible, raising it as a question by threatening a senselessness in relation to which sense can be restored. Rationalism rejects context in favor of the binary relation of cause and effect, such that sense never becomes a question and need never be made, only identified.

Instrumentalism is also unsustainable in a contingent world, for it too relies on a stable set of relations and a consistent set of rules. Contingency renders all relation unreliable, as it may invoke a new rule to trump or recast the standing ones, decoupling chosen means and desired ends. Still caught in a web of relations, means under contingency assume an antipositive autonomy, suspending or superseding the rules that would tether them to predictable ends. Contingency frees the world to go where it will, to choose another path, such that its cooperation is never guaranteed, setting awry the best-laid plans.

The digital also demonstrates a certain kind of freedom, but the difference between the digital's freedom and the freedom of contingency is telling. Contingency frees the world to choose a new path, a choice that is neither rule-bound nor arbitrary, but is negotiated within all the relevant relations, all the threads that tie the situation to itself and to its context. The digital's freedom, by contrast, is always a choice between 0 and 1, which have only a formal relation. There is no inherent relationship in the digital between a thing and its context, because, inasmuch as they are connected, that connection is effectively arbitrary, determined by a set of rules that have been deliberately assigned and can be as easily altered. Within a given digital object, internal relations are similarly tenuous; individuated and without any evolved or inherent relation, a thing's qualities, its parts and their characters, can be altered at will, leaving all else about the thing untouched. As we will see in the chapters to come, digital things have no essential integrity, no reason for being how they are, and are thus exposed to arbitrary manipulation. Nothing holds together a digital thing, nothing makes it what it is, so that its freedom is only the meaningless freedom of formal alteration without consequence. The digital is unrooted, decontextualized, individuated, and so can never really make sense, for its reasons are mere mechanisms, calculations that of themselves mean only logic and arithmetic and nothing more. The digital world is indifferent to itself, whereas the actual engages itself reflexively in an unbounded depth and tangled complexity.

Thus the digital has already been disqualified as a domain of sense-making, for in the digital, the progress of each step to the next is inexorable and necessary. The logical calculations that determine what step is next in a digital process can generate only one possible outcome, the correct one, and the discreteness of the digital ensures that this one outcome has no margin of uncertainty, no fuzziness, no ambiguity that might afford contingency an opportunity to exert its playful or tragic disruption. There can be no sense but the vacant sense of 0 and 1 in the digital, for there is no nonsense either, no question that impels a rescue of sense from its dissolution.

Does this speculative metaphysics end up reproducing the very binarity it aims to supplant? Is the mesh simply the contrary of the posit and contingency only binary's opposite number, such that this metaphysics that begins with productive difference turns out to rest on the familiar bed of identity? Binarity is indeed at odds with contingency, yet the two metaphysics are not opposed as binaries. Rather, contingency's creativity extends even so far as to produce its own resistance. It is everywhere, a mesh that churns out knots and whorls, patterns and consistencies, to generate semistable communications, simultaneously constructing and assailing the things that would be posits. The mesh congeals (or fails to congeal) in all kinds of patterns, some dense, some diffuse, some chaotic, some ordered, and where there arises regularity, there is the beginning of a posit. This ontogenesis does not finally arrive at a wholesale positivism, for positivism requires a genuine reversal in which the things, the protoposits, would claim a priority over their relations. Instead, on the mesh, there always remains a degree of instability, a connection from the semistable knot or complementary community to its outside, to all of the outside. This would-be-thing, a persistence in the fabric of the mesh, may trail off into its milieu, or it may draw boundaries around itself, but in any case, it is not yet cut off from its outside and without relation. One could understand such actual things as *the seeds of posits,* but only if one remembers to preserve the germinal or embryonic that is never exhausted even as the seed grows in every direction.

Crucially, then, the rudiments of positivism are as ubiquitous as the contingency that gives rise to them, admitting order without elevating it to a principle. It is contingency that lays the foundation for a regime of positivism, though it does not follow through, stopping short of the priority of identity. This explains the possibility of positivism but also demonstrates how the contingent world of change and motion can include definite *things*: objects, selves, distinction, even identity, a whole panoply of protoposits with more and less discrete borders, more and less stable relations, a concerted insistence of selfhood in some cases and a promiscuous preparedness to meld or melt in others. Warding off a binary between contingency and the positive, there are all sorts of in-betweens, degrees of consistency, a vast, occupied space between the molecular and the molar.

Thus does the digital take hold even without its designated technologies: the world of contingency makes ready for the abstraction of identity, the conceptual slice that distinguishes absolutely between one thing and another. Contingency is everywhere, but likewise, in its own way, is positivism everywhere, awaiting its apotheosis in the technologies of the digital. Contingency's power of creativity extends even so far as resistance to contingency.

Ontology of the Digital

The ontology of the actual locates contingency at the heart of the world, contingency as agent of creativity and change, but also as instigator of (partial) identity and (finite) stasis. The ontology of the digital shows a marked contrast with that of the actual: digital ontology carves out, amidst the actual contingent world, its own world of *necessity,* which explains both the digital's irredeemable deficit but also its remarkable capacities and its wild appeal. For a necessary world speaks to our ardent desires, offering immense satisfactions and even answering to some of the most persistent questions at the core of human being. How do digital technologies expunge the contingent and institute this regime of necessity?

Contingency wraps the world in its meshwork, weaving things into and out of a complex web of relation, holding the contents of the world in place and also freeing those contents for new adventure. Nothing in a contingent world is entirely alone or independent, for everything is enmeshed, everything touches everything else: *con-tingency,* "touching together." To escape contingency, the digital builds its world from a universal element that, from the outset, exists without relation, in touch with nothing. That building block, the bit, stands alone, the archetypical posit, bearing no meaning but that of its value. And even the values of a bit are formalities with no substantive meanings of their own; to label those values 1 and 0 (or, in some contexts, *true* and *false*) is a widely accepted convenience, making it easier to talk about bits, but those labels imply a numerical (or logical) significance that bits themselves do not carry. Not inherently numerical, a bit's values signify only the values' formal distinction from each other, *this* and *that,* with 0 meaning nothing more than "0 and not 1" (or "*this* and not *that*"), while 1 means only "1 and not 0."

The digital wards off contingency by building its entire world out of these simple, unambiguous binary digits. Because bits are discrete and exact, because they are patently simple, they include no ambiguity or uncertainty, leaving no room for accident or indeterminacy. Unlike the actual world, which is buffeted by contingency, bits are just what they are: isolated, autonomous, positive individuals. They are bound by no relation and offer no resistance, exhibiting a stark clarity that affords contingency no purchase. Shorn of the meshwork of relation that infuses all of the actual, the digital is sequestered in its own world, a world of immense possibility, ready to admit everything that takes the form of information. Because the bit has no meaning of its own, because even its values signify only a formal distinction from each other, the bit is neutral and indifferent, available to take on any assigned meaning, any meaning that can be encoded in 0s and 1s.

Everything that happens in the digital, all the data and all the instructions, the input and the output, the internal calculations and the worldwide transmissions, text, image, sound, icon, menu, window, cursor, app, algorithm, website—all of it is represented in the machine as sequences of 0s and 1s, the only two terms in the native vocabulary of the digital machine. That only two values represent all the operations and all the data that pass through digital devices might seem implausible, but that is why there need to be so many bits, even for relatively simple operations. The elemental action of the computer by which it carries out all of its operations, no matter how many billions of steps are involved, is a comparison between two bits to yield a single-bit result. As Figure 1 illustrates, there are sixteen different operations that take two bits of input and yield a single bit of output. Each of these sixteen operators, often called "logic gates," must give a result for each of the four combinations of values for a pair of bits: 0,0; 1,0; 0,1; and 1,1. The AND gate, for example, takes a pair of bits as input, and outputs 1 if both of the input bits are 1, and otherwise it outputs 0. For the most part, this logical AND matches our usual intuition about what *and* means: we say that "x and y" is true only if both of them, x and y, are true. The OR gate outputs 1 if either (or both) of the input bits are 1, else it outputs 0. (In other words, "*x* or *y*" is true whenever either of them is true [or both].) Everything the digital does, including arithmetic, is an application of staggeringly long sequences of these sixteen gates.

The Sixteen Binary Logic Gates

AND			OR			ONLY-IF			XOR		
0	0	0	0	0	0	0	0	1	0	0	0
0	1	0	0	1	1	0	1	1	0	1	1
1	0	0	1	0	1	1	0	0	1	0	1
1	1	1	1	1	1	1	1	1	1	1	0

NAND			NOR			NOT-ONLYIF			XNOR		
0	0	1	0	0	1	0	0	0	0	0	1
0	1	1	0	1	0	0	1	0	0	1	0
1	0	1	1	0	0	1	0	1	1	0	0
1	1	0	1	1	0	1	1	0	1	1	1

						NOT-X			NOT-Y		
0	0	1	0	0	0	0	0	1	0	0	1
0	1	0	0	1	1	0	1	1	0	1	0
1	0	1	1	0	0	1	0	0	1	0	1
1	1	1	1	1	0	1	1	0	1	1	0

X			Y			TRUE			FALSE		
0	0	0	0	0	0	0	0	1	0	0	0
0	1	0	0	1	1	0	1	1	0	1	0
1	0	1	1	0	0	1	0	1	1	0	0
1	1	1	1	1	1	1	1	1	1	1	0

Figure 1. In each of the sixteen gates in the table, the first two columns represent the input values of two bits (*x* and *y*), and the third column is the output value. For example, the third row of the OR gate indicates that an *x*-value of 1 and a *y*-value of 0 yields an output value of 1. Two of the gates are unlabeled because they don't have standard names. The last six gates are not formally binary operators; the final two are nil-ary operators because their output values don't depend on any inputs, and the other four (of those last six) are unary operators because their output values depend on only one of the two inputs.

Digital operation is thus the concatenation of billions or trillions of trivially simple calculations. There is no room in this picture of digital operation for fuzziness, uncertainty, or indeterminacy: specify one of the sixteen gates along with the two input values, and the single output value is clear and inexorable. Bits are always either 0 or 1, and calculations on them yield an unambiguous result of 0 or 1; such calculations, tantamount to looking up a value in one of the sixteen tables, are never obscure or problematic or strange. The system is straightforwardly deterministic. No matter how many calculations it takes, a digital action is always just a sequence of simple table look-ups where a pair of bit values is referred to a particular truth table (another name for a logic gate) to return the one and only possible result.[1]

This description of digital operation is not an abstraction or figurative simplification. Bits and the logic gates that operate on them must be materially realized in any digital computing device, and while different materials could be (and have been) used, the core hardware of a digital device today is a silicon chip, a small wafer, on the surface of which are etched circuits, pathways along which electricity flows. Within a computer's processing chip, or traveling to and from storage devices or computer memory, bits are realized materially as the voltage values of those flows of electricity; within a given system, one voltage value is assigned to signify 0, and another voltage is assigned to mean 1. For example, many computing systems use a value of +2 volts to signify (a bit value of) 0 and +5 volts for 1. This convention of electrical voltage values coursing through etched pathways on a chip's surface is how bits flow through real digital machines.

Bits also stand still in digital machines, housed briefly in memory registers on chips, enduring a bit longer in random-access memory (RAM), and persisting for the long haul on hard drives, compact discs (CDs), read-only memory (ROM), flash drives, and other digital storage media. The principle of bit storage is more or less the same in each case, but the details vary depending on the properties of the physical substrate in which bits are to be written and read back. Many media, such as a conventional hard disk, encode bits magnetically: a magnetically receptive surface is divided into circular bands (or a tightly wound spiral, like a vinyl record), and these bands are themselves divided into tiny spots; each spot of magnetic material generates a magnetic field oriented in one of two possible directions. These fields can be both measured and altered (i.e., read and written) by magnets passing above that surface. Similarly, CDs store 0s and 1s also by dividing their surfaces into tiny spots, each of which reflects light strongly or weakly, hence the distinctive rainbow shimmer of a CD struck by light.[2]

Summarizing: the 0s and 1s of bits are instantiated in material media by assigning two measurably distinct states of some property of the physical medium to stand for the values 0 and 1. But this materialization of bits using physical properties of a substrate presents a fundamental problem: the digital's operation, its construction of a world of necessity, depends on bits behaving perfectly, occupying a state that is either exactly 0 or exactly 1; but, because materiality is part of the actual world, physical properties are generally inexact and not entirely stable. Magnetic field orientations, laser beam reflection times, and electrical voltages are all haunted by the inherent contingency of materiality; an electrical potential will never measure *exactly* five volts, but will only approximate that ideal value to some degree of precision. How do the imperfect values as measured in an actual computing machine, which are never exactly equal to the nominal values assigned to 0 and 1, represent the perfect 0s and 1s that are the hallmark of bits and the bulwark against the contingent?

A brilliant engineering innovation addresses this mismatch between the materiality of the bit and its idealized value, defying the flux of contingent materiality to constitute actual, physical bits as perfect, logical specimens. Through this clever but simple solution, digitalness becomes materialized as technology, such that the nonpositive continuum of nature serves the perfect, discrete positivism of the digital. Here's the trick: material measurements *close to* the nominal, ideal value are treated as though they were *exactly* that nominal, ideal value. In a system where a value of five volts has been designated to represent a bit value of 1, as long as the electrical potential of a given materialized bit is near 5V, the system (of silicon circuits) will produce results associated with a logical bit value of *exactly* 1. For example, at a given instant, an electrical current entering a logic gate as an input might measure 4.8V, but the gate is physically designed so that it still performs the correct *logical* calculation, treating this input as exactly a logical 1, and outputting an electrical potential that will be treated at the next logic gate in the circuit, again, exactly as a 0 or 1. (As discussed above, whether a binary gate outputs a 0 or a 1 depends on the other input value and on which of the sixteen possible gates it is.) Thus does the lowest level of digital operation construct bits *to act in practice as their own ideals*; the bit is a materialized abstraction, perfectly equal to its assigned value, even as it is concretized in imperfect and imprecise materials. And this coup of digital engineering, which underlies so much of the digital's power and reliability, explains how the digital discounts contingency. It removes contingency by treating every bit as an ideal, exact value, where accident, becoming, disidentification, and error are virtually eliminated, leaving no play of potential, but only the rigid rule of necessary determination. This defeat of contingency—to admit and then ignore small deviations from the ideal—is the ingenious maneuver that avails technology of the awesome power of the digital, rendering that technology immeasurably useful by bringing abstraction into the concrete.

Retaining the bit's abstraction even in its concrete actualization, leveraging a discrete and self-equivalent element as foundation, digital technologies eschew the messy and irregular materiality that ensures the limited reliability of nondigital technologies. Constructed out of this perfect, invariant element, the digital machine operates with mind-boggling reliability, electrical rapidity, and increasingly minimal demands of space and resources. To do so, it erects another world in which to travel without friction, an incorruptible world, where the altered ontology neuters the uncertainty of being and stills the constant churn of becoming, replacing it with the icy perfection of individual and exact steps, a single possible outcome and no room for error. The digital summons to earth the heaven of Platonic forms, proceeding without flaw and barely tethered to its own materiality, operating in a different world, a digital world. Untrammeled by the bonds of matter that burden (but also enable) so many technologies, the digital becomes weightless, without resistance, a rarefied technology of pure logic. The payoff of this technological marvel is the amazing power of the digital, all the extraordinary things the digital can do. So, again, *what does the digital do?*

The Lion King

A trailer for the 2019 Disney "photorealistic" remake of *The Lion King* film shows the wind, digitally animated, blowing through the mane of an adult lion. Is this wind exposed to contingency, influenced by a falling tree, or by the radiant heat rising from the ground as evening approaches? In fact, in the digital encoding of the animated world, there is no wind, not even a simulated wind, but only an algorithm to simulate the *appearance* of wind passing through fur. And this wind-as-waving-fur adheres strictly to a mathematical formula that models the shifting texture of blowing fur, with no influence from factors outside that algorithm. Execute it a second time and the exact same pattern would recur; without contingency, the digital wind remains loyal to a rule, a formula, such that nothing weird can blow. This allows the animators an absolute control, for they are the algorithm's authors and can tweak it however they choose, down to the smallest detail, perhaps to enhance the ennobling quality of the windblown mane.

Minimizing contingency, the digital can offer only a muted creativity; it can produce nothing truly new, but only rearrange the forms it is given. Nothing in the digital world admits the unexpected, the breathtaking surprise of a sudden gale. The urgent question that drives this book is whether this trade-off is worth it: do we wish to sacrifice the richness and spontaneity of our all too human world to gain instead a world of perfect compliance and universal reason? But it isn't clear that this question gains any traction in the digital age. For digital techniques are so beguiling and its technologies so widespread that we already see the world from a digital perspective and judge the world according to digital values. The digital is a positivism, encouraging those under its influence to credit only posits. The digital is a rationalism, satisfying but also stoking the desire for a world that makes sense, always and everywhere. Thus does the digital preempt critique, erasing from our worldview the contingency that we no longer recognize as missing. Under the sway of rationalism and positivism, contingency cannot appear except as an impediment to efficient progress, which casts the digital's detractors as carping cynics or crusty luddites, unable to appreciate the great value of a nature finally subject to a plenary rational order.

By itself, a bit cannot do much. A single bit can fully encode (a conceptual space of) two things; we assign 0 to refer to one of them and 1 to refer to the other, and then we indicate one or the other of the two things using a single bit. Adding more bits, one can encode a greater number of distinct things, but this also adds another distinction, beyond the distinction between 0 and 1. As soon as there is more than one bit, there is also a discrete distinction between individual bits. Bits are typically treated in sequence, such that each has a *value*, 0 or 1, and each also has a *place* in the sequence, the nth place. These two complementary differences allow bits to work together to encode complex structures. A sequenced group of n bits can represent 2^n different values, and so can encode a space of 2^n different things. There is a natural encoding in which sequences of binary digits (bits) represent positive integer numbers, just as we usually represent those numbers using sequences of decimal digits: a single decimal digit has ten possible values, 0 through 9, and so can encode ten possible things; a sequence of four decimal digits can encode 10^4 (10,000) different things, such as the numbers 0000 (or just 0) through 9999. Likewise, five sequenced bits can represent 2^5 (32) possible values, corresponding to the binary numbers 00000 through 11111. With two discrete distinctions, value and place, bits can encode all information.

As has been emphasized, all data and all processes in the digital are represented using codes of bits, sequences of 0s and 1s. Data are not just factoids about people and objects in the world, nor are they just what one inputs as content or adjusts as parameters when using an application; everything in the computer, even the windows, the icons, the desktop background, or the mouse location, anything that the digital takes as an object for analysis, manipulation, or reporting—every *thing* in the digital is a sequence of bits that works as a code.

One might similarly characterize alphabetic or even ideographic writing: a code of discrete values, often used for information storage, retrieval, and distribution. And like bits, symbols such as letters and numerals work by instantiating an abstract form, independent to some degree from the material substrate that hosts it. The letter k appearing in a written (or printed) word is, in this sense, a perfect k, in that it functions by being recognized and then understood as precisely this letter, not an approximation of a *k* or almost a *k*, but exactly *k*, equivalent (in that regard) to every other instance of the letter *k*. This is not to say that every *k* looks identical to or is equally as legible as every other. But, even if the ink is smudged or occluded, and even if the font is heavily stylized, the letter form still functions perfectly as long as it is accurately recognizable. While this demonstrates that the innovation behind the bit is not an original stroke (letters, too, function as their own ideals), the bit differs from letters in that the idealization is externalized, taking place in the machine rather than in the human act of recognition.[3]

Bits thereby obtain a power that exceeds, at least in one respect, that of writing: the bit exhibits a duality wherein it becomes *writing that does what it says*. The bit is both an electrical current (or magnetic field orientation, etc.) that determines the electrical output of logic gates to make a digital machine operate but is also and simultaneously a symbol (0 or 1) that signifies, as part of a sequence of bits, by virtue of a code. This conjunction of doing and saying reveals another aspect of the bit's awesome power that in turn empowers digital technologies. Because bits both do what they do and say what they do, digital devices can be automated, given instructions that can effectuate their own operation.

The fusion of saying and doing is most evident in the low-level code of bit sequences known as "machine language," the "language" understood directly by the central processing unit (CPU) of a digital machine. As different digital devices use different CPUs, there are many different machine languages, different codes and different commands for each type of CPU. Bit sequences encode machine language commands in a manner somewhat analogous to the way that letter sequences form words. In the machine's native language, a sequence of bits encodes a command more involved than the simple logical comparisons (gates) that deal only with two bits at a time, but machine language commands remain fairly basic instructions. A typical command of machine language might instruct the CPU to increment-the-value-of-the-number-currently-stored-in-the-primary-memory-register or to move-the-data-in-memory-section-A-to-memory-section-B, where A and B are numbers that specify locations within the computer's numerically indexed memory. The examples just given might be abbreviated INCR or MOVE, respectively, and such commands instruct the CPU to carry out an operation of arithmetic or logic, such that a person who knows the sequence of bits for a given command can look at that sequence of bits and understand what the command *says*.[4] But, as a sequence of bits, the command also enters the CPU as a collection of parallel electrical signals, one for each bit in the sequence,[5] and the pathways that these signals follow across the surface of the chip, along its channels and through its logic gates, generate further electrical flows that effectively accomplish what is commanded (what the command *does*). The same bits that mean increment-the-value-currently-stored-in-primary-memory also materially cause the value currently stored in primary memory to be incremented (increased by one). It's as though the letters in the printed phrase "walk the dog" could somehow rise out of the page (or screen), leash her up, and take Mittens out for her constitutional.[6]

Machine language is one example of how sequences of bits are used to make a code, a code of CPU commands. But bit sequences encode everything in the digital machine, including internal data that would have little meaning for a nonprogrammer, and also all sorts of things that appear in the human–computer interface. A scheme for encoding colors might use a sequence of thirty-two bits to indicate a particular color, while the venerable (but still widely employed) ASCII code (American Standard Code for Information Interchange) uses sequences of seven bits to represent typewriter characters. For example, the ASCII code represents our friend the lowercase k using the seven-bit sequence 1101011, which is the familiar decimal number 107, and ASCII represents a plus sign (+) with 0101011, or decimal 43. As mentioned above, a sequence of n bits can represent 2^n possible different values. Thus, the seven-bit ASCII code has 2^7 (128) different code values, each of which corresponds to a different typewriter character (including not just uppercase and lowercase letters, but also punctuation marks, line feed, backspace, and spacebar, for example), while a color code with a thirty-two-bit depth can encode 2^{32} (or about 4.3 billion) different shades.

Another code might use two sequences of bits to specify a pair of Cartesian coordinates, one sequence for a horizontal index (often called an x coordinate) and a second for a vertical index (y).

Add a third sequence that uses the thirty-two-bit color code just referenced, and one can encode an entire image with a list of these threesomes of bit sequences. Such pixel-based representation of visual information—an image as a collection of colored dots in a two-dimensional grid—is even built into the hardware of our digital machines: computer and smartphone screens are controlled by a stream of bit sequences that specify which pixels should be which colors.

All commands and all data in a digital machine are made of sequences of bits, including everything that shows up in the interface but also everything the user inputs into the machine. The user slides the mouse or strikes a key or touches the tablet, and this action generates an electrical signal that passes into the machine and is interpreted as a sequence of bits representing a value that corresponds to the user's action, containing information about which direction and how quickly the mouse moved or where on the screen the user touched or which key was pressed and for how long. These too are codes, input gestures as sequences of bits that effectively become commands for the CPU. They illustrate again the duality of doing and saying; the bit sequence generated when a key is pressed is determined by a code that maps keys to sequences of bits, but that same bit sequence is read into the operating system as it executes and becomes an instruction that results in an action, such as making a letter appear on the screen.

Slot Machines

Las Vegas casinos have replaced their old slot machines, which had real rotating reels, with digital simulations. Pulling the lever of the old machines transferred rotational force to the three cylinders, which were slowed by a gentle resistance until each came to rest at a détente in the gearing mechanisms. The signature lever of the "one-armed bandit" has not entirely disappeared, but where it has not been replaced by a button, it now triggers an algorithm that cycles through a sequence of symbols in each slot until finally settling on some algorithmically determined resting combination of symbols. The rotation of the reels in the old machines is an essential part of the process, an injection of contingency, for the physical system of mild resistance and angular momentum is highly sensitive to tiny variations in initial (and contextual) conditions, such that each spin has a complexity beyond calculation. Digital machines, on the other hand, include only a simulated contingency: to determine which symbols appear on the screen, they use "pseudorandom" number generators, which *simulate* randomness by applying a formula to the internal clock of the machine (or some other suitably unpredictable and changing value).[7] The formula is designed to yield nonlinear results, with a heavy influence from the least significant (most precise) parts of the digital mechanism's internal digital clock. Nonlinearity helps to ensure that two pseudorandom numbers generated in quick succession will still show a fairly high divergence from each other, which makes the sequence of numbers (hence also the sequence of symbols on the machine) hard to guess.

Either of the two slot machine systems, analog or digital, could be unfairly designed, favoring certain outcomes while presenting the appearance of impartial selection. Extra weight placed asymmetrically around a physical reel will make it more likely to stop at certain symbols. And on a digital machine, it is a simple matter to code a bias into the algorithm, so that some symbols, or even some combinations of symbols, are more likely to be selected than others. There is therefore no reason to think that the physical reels are more fair than the digital ones.

But even if the outcomes of two different machines, one digital and one analog, are statistically comparable, the sense of play is not. For one thing, on a digital machine, the sequence of symbols that precedes the final one on the screen is entirely superfluous; the resting state of the digital machine bears no inherent relation to the symbols that flashed across the screen prior to its stopping point. Unlike physical reels, there is no sense in which the immediately preceding symbol on a digital machine was *almost* chosen.

But the bigger (though hidden) difference between the two cases is the nature of the randomness involved. In the older machine, each spin of the reels is an appeal to the whole universe, to all the forces, however distant, that play a role in the highly sensitive system of friction and momentum. Anything might be relevant, from the temperature that day, to the gambler's arm that is sore from carrying suitcases, to the position of the moon.

By contrast, the digital system is determined by rules, rules that are absolute, that generate a particular symbol as a result of particular inputs; such a system is insensitive to the force with which the lever is pulled, to the air pressure, to the age of the machine, and to the contingent productivity of the world. The digital is sealed off from its outside, occupying a positivistic and deterministic world of its own, which might seem like an unlikely place to gamble on a slot machine.

This also explains why a digital *I Ching* would be an absurdity (which is not to say that such systems do not exist). One tosses the yarrow stalks as a way of entrusting the universe to direct one's attention. This method of seeking guidance rests on the metaphysical supposition that things in general are connected, that the factors that scatter the stalks in a pattern, which generates a hexagram that designates a passage in the text of the *I Ching*, are the same factors presently at play in one's life, and that those same factors shape the interpretation of the *I Ching* text. Though the method of the *I Ching* uses six binary choices (yarrow stalks or coins or whatever is available) to generate one of sixty-four possible hexagrams, the interpretive gesture essential to the method ensures the preservation of contingency from the initial throw to the subsequent thoughts and actions. A more reductive and less interesting reading would allow that a digital *I Ching* is just as good as yarrow stalks because the selection of a passage need only be (pseudo-)random, and not necessarily connected to the whole universe; even a randomized selection suffices to point to a textual passage that leaves room for the questioner's unconscious to invent a meaning out of the encounter of that text and her self.

No matter how many bits one strings together, and no matter how extensive the encoding schema that those bits serve, the sequence itself, any sequence of bits, retains the underlying characteristics of an individual bit: discreteness, self-identity, neutrality. Contingency gains no footing in the digital through the multiplication of bits; adding the distinction of place to the distinction of value still yields a pure positivism. Digital ideology discourages or even disallows the creative foment of contingency, but rejecting contingency, it gains the clarity of posits, the order of an assured reason, and the ready utility of the instrument. This trade-off, sacrificing the challenge and excitement of surprise for a flexible and useful regularity, is the formula of the digital, the overarching strategy of computation; predictable, determinate, ordered, and manipulable, the computer minimizes its resistance to maximize its availability. But where's the fun in that? Digital ideology may make computers astonishingly capable, but it would seem to make for a much less interesting digital world. Whence the appeal of positivism, rationalism, and instrumentalism?

There is indeed something immensely comforting about a world where posits have primacy, where everything makes sense, where calculated intervention reliably leads to the satisfaction of one's desire. A previous section on digital gaming cited Christopher Douglas's theory that some games are pleasurable ("existentially soothing") precisely because they locate the player in a world designed specifically for her, a world in which every object has a reason in relation to the player, and every relevant task is achievable. The game world includes not just the means to succeed, but even the ends that signify success; the game thus provides both desire and its satisfaction, prescribed and achievable goals that the player adopts in order to play the game. (Notably, there are also practices of countergaming that flout the expectations built in to the game, but even those typically depend on the recognition of what the game asks of the player.) Douglas's analysis focuses specifically on certain genres of gaming, but the structure of desire and satisfaction that he identifies is built into digital media most generally; games provide an archetype, a perspicacious example of a way of being that applies to the digital in its entirety.

Douglas's theory of "existential soothing" helps to resolve one of the core paradoxes of digital gaming: if the activity of gaming generally is about exercising a certain freedom, the freedom of *play,* then how can we experience digital games as satisfying, given that the digital excludes the contingency that would support such playful freedom?[8] Digital games amount to a set of strict rules wherein the goals have been established and the means to achieve them preprogrammed. Instead of an exercise of experimental and inventive freedom to forge her own path, the player must discover the path already laid out by the game's designers and programmers. Existential soothing offers a narcissistic compensation for this strangled freedom: the player does not get to do what she will, but by accepting the severe limitations on her actions in the game, she gains the immense satisfaction of a world perfectly suited to her capacities and even designed so that she is at its center. Everything in the game is put there for the player, who may be playing a character narratively coded as abject or insignificant, but who can nevertheless feel herself in the game as the central subject and referent of game design.

A first-person-shooter (FPS) game typically allows the player to move forward and backward, to move side to side, and to turn continuously in a circle. The player's avatar may be able to duck into a crouch and jump to clear hurdles or climb up to a ledge. There are commands that instruct the avatar to fire a weapon, maybe change to an alternate weapon, reload ammunition, and aim down the gun's sights. This whole set of actions for the player is designed to make possible the game's central goal, to move through a simulated space shooting at simulated enemies. There may be other possible actions, but moving and shooting are the basic capacities of the player in an FPS game. Many such games also emphasize a kind of visual realism, which is widely thought to enhance the intensity of the game, making it easier for the player to set aside, at least partially, her awareness that it is only a simulation.

The question is why this realism would be at all satisfying when the sense of embodiment and the freedom that accompanies embodiment are so thoroughly restricted. The player may be able to peek around corners or wave to other players, but may not be able to look at her own feet, or shake hands with another player's avatar, or skip, or throw dust in the air. In some games, the player can't even walk, moving only by running wherever she goes. One might expect the limited palette of bodily options to impose a strong sense of unfreedom; in a game where movement is a primary capacity and a basic means of achieving the game's goals, the restrictions on bodily movement ought to be acutely frustrating. But for most players, the logic of play works in the other direction: only when the player embraces the limitations on bodily capacity (and other possible actions) does the game become challenging enough to feel like a test (or ego-syntonic exercise) of one's ability. Only because the player is aware at some level that her range of freedoms is closely circumscribed does she feel a strong sense of accomplishment when completing a designated game goal. The game world and its predesignated goals are designed not to allow a maximal freedom of play but to complement exactly the limited affordances of the avatar.

This analysis, built on Douglas's notion of existential soothing, extends well beyond gaming to describe the user's relationship to digital machines generally. It's not just getting what one wants in the digital; the whole digital world is a construct of what one wants, a world designed both to pique our desires and to satisfy them. Almost any application on the phone or computer lays out the available options, and any chosen option is then exercised at once: desire and its satisfaction. Similarly, this book locates the digital ideology as both the condition and the consequence of the massive popularity of digital technologies. Digital ideology prepares us to prefer a digital world, wherein we are trained to formulate desire in digital terms, such that we relate to the whole world, even outside of the computer or phone, in a manner akin to the way we engage with digital machines. The consequence is that contingency goes unmissed, its absence celebrated as an increased efficiency, predictability, regularity, and compliance.

Politics and Positivism

One might suppose that the national politics of the United States today, having abandoned sense in favor of purely phatic performance, would elude digital capture, for the digital's dominion does not extend to nonsense. The nonsense word, as Gilles Deleuze points out in *The Logic of Sense*, is the only word that speaks its own sense; it *is* nonsense and it *means* nonsense. The computer can neither make nonsense nor account for it; each processed word triggers some subroutine, and the computer is indifferent as to whether that subroutine performs a complex action signaling that it has correctly parsed the word or a null operation that effectively ignores the word. One subroutine is as good as another, and the computer can, at best, decide that a given word is not among the words in its vocabulary. But nonsense cannot be judged by inclusion in a digital dictionary, for it is not what stands outside of sense, but rather a kind of pretense of sense that then fails to deliver the goods. Nonsense, in the digital, would be the accommodation of what cannot be processed, the computation of what does not compute.

But in actuality, today's politics are a vicious and hyperbolic assertion of the values associated with the digital. They are an instrumental politics, where any aim has been replaced by the act of pure instrumentalization. At least on the big stages, politicking means treating people (and everything else) as means rather than ends, such that ends fall away and means without aim become the currency of the age. Money once offered the perfect medium of auto-instrumentalization, because it is precisely a means that has eliminated its end, a means unto itself. (The only aim is to make more money.) But the digital catalyzes an even more universal solvent, swallowing not just goods, not just capital, but all information, everything that's anything, into its endless chant of 0 and 1, those meaningless tokens that now preside over all the world's meanings. Politics practices a thoroughgoing rationalism, a parody of reason that leverages a reason made frail by its confinement to binary answers paired with depthless questions. Reasons are still sometimes on offer, but reason stripped of its spontaneous and vital difference becomes nothing more than taking a side, a rhetorical battle whose only winner can be the abdication of sense, even the denial of sense. Politics today makes no claim to sense, which is now castigated as the tool of the oppressor, who is portrayed as so neurasthenic as to appropriate (what was once) the priest's source of power, a discourse instead of a proper fight. Politics attempts not to appease, but to stoke *ressentiment,* a dangerous weapon for sure, but immensely powerful nevertheless.

Finally, politics trumpets positivism and can no longer be distinguished from that positivism. Winning is victory, victory for the positivist, but also the victory *of* positivism. It's not just that winning has become the end, the end that is its own satisfaction, a pure instrumentalization. It is moreover that the message of a bottomless positivism, a positivism that also is an end unto itself, is what enables all of the other vacancies of value. Positivism, reducing everything to a binary that has no reference, no bottom line, vacates the significance, the force of all values, leaving only a specious bickering in its wake. It is precisely the insistence on positivism that makes it possible to assert that 0 is every bit as legitimate as 1, that truth has no ground but the rhetorical détente of a two-sided opposition.

It is weak evidence of the digitalization of politics, but evidence all the same, that many recent national elections have arrived at a stalemate. (*Digitalization*, in this case, refers not only to the use of digital tools to target voters, disseminate messages, organize campaigns, etc. It refers more essentially to the submission of politics to the values of the digital, the positivism, rationalism, and instrumentalism that both buoy and enjoy the digital's ascendancy.) Most famously, the 2000 presidential race between George W. Bush and Al Gore pitted equipotent calculating mechanisms against each other to split the country, and the state of Florida, right down the middle. The result was an election decided by lawyers and a handful of judges, who acted all too predictably in accord with principles of loyalty to party rather than to justice. If the stalemate left the election to the play of contingency, then this was a contingency battered into submission, ready to sell out to the highest bidder, a contingency neutered and without decisive force.

Contingency fills out the middle space, it serves as a bulwark against positivism, ensuring that there is always another option, a third way. To erode contingency and to erase it even from our cultural and personal memories is to open the path to a politics that includes only independent, positive positions, bearing no complex or nuanced relationships to each other, such that one can choose only one or the other option. If American politics has always tended toward the most extreme positivism of a two-party system, the tragedy of that meager nuance has become all too evident in the digital age. As usual, the immediate consequence of digital politics—in this case examples include digital advertising and calculation resulting in victories that are only statistical variation from a stalemate—also includes a deeper illness, the extreme polarization of society that empties out the very possibility of what used to be called politics. It's not just that people now simply choose a side and don't look back, as though that were the activity and responsibility of politics; it's that they don't even imagine that politics might be something else, that it might be discourse, ethics, sacrifice, practice, personal. Simply to choose a side is to abandon the responsibility of politics, a responsibility whose telos or finality is voting but whose body is far more and much else.

The politicians, or at least the machinery that props them up, have figured out that the middle space, the space of contingency, has been thoroughly hollowed out. And without that middle plane, confronted by only positive, individuated, disconnected points, truth has no ground and can be manipulated *ad lib.* When the idea of truth, or the concept of a fact, is become an isolated point, a claim whose only meaning is to be true or false, then the Orwellian truth, the irony of truth of irony as truth, has as much legitimacy as any other truth claim. (*True* and *false* have no reference, no meaning, aside from their contribution to which side one picks in the empty battle of political choice.) That is, with contingency absent, it cannot defend itself against the charge, leveled at the same time as condemned, of a nihilistic destruction of truth. Contingency as the middle ground, the nuance between truth and falsehood, is retrospectively diagnosed as an insult to positivist truth, a destruction of truth, when in reality (and by contrast) it is the bald assertion of truth without relation that undermines the force of truth and renders it fungible.

This sort of analysis is more provocative than satisfying. One suspects that it would be just as easy to offer a reading of recent American presidential elections or politicized popular rhetoric that revealed amidst the apparent polarization a significant complexity, a vast middle space, occupied by our huge heterogeneous population, but betrayed by the statistical analysis that erases the fuzzy middle in favor of the numerical confidence of the two poles (as evinced by many polls). One hears plenty of arguments to the effect that the traditional poles of American politics, represented more or less by the two dominant parties, have broken down, that the political landscape has shifted, that simple attempts to group voters by class, race, educational achievement, recency of immigration, and so on, all fail to explain or predict political allegiance. In other words, many people feel that the political situation has become unaccountable, too granular to synthesize into any kind of accurate sense, as though the crude simplicity of left and right has ceded to an indefinitely nuanced individuality.

Surely both descriptions are valid: the very idea of nuance has been kicked to the curb, but this still leaves room for a great many distinct positions, each constituted by a selection of binary (or, in any case, discrete) choices on different issues. The elimination of nuance, the spite for the hesitation that occupies the middle ground, also eliminates the fabric, the mesh, that ties issues together and that connects the two poles (rather than simply leaving each as a posit), and so makes room for a menu-driven identity, a cherry-picked, customized position whose distinctness is a matter of statistically measurable differences from any single other position, but which is still constituted as an aggregate of individual molar positions. (It is at this point that the statisticians step in to offer *nearest neighbor* analyses and identify the most distinctive categories, or topics, by which to group voters, yielding a new truth based not on sense, but on the objective validity of numbers. It isn't clear what we're supposed to do with this truth, but the campaigners seem to have figured out how to leverage these categories for precisely targeted propaganda.) This is the image of identity in the age of positivism, identity as aggregate adherence to conceptually definite categories.

The practice of realpolitik around the world no longer even genuflects before the classical transcendentals that formerly served as a kind of final (if violable) limit on behavior and policy. There is now ample reason to believe that Donald Trump really could stomp down Manhattan's Fifth Avenue, machine gun in hand, laying waste to the pitiable human beings in his path, and his supporters, at least, would celebrate his boldness and manifest moral authority. Realpolitik does not require the erosion of contingency to get away with murder, but it benefits from the derelativized, an-ethical playing field left behind when the middle ground of contingency is washed away, a field of disconnected posits that, without relation, are all at the same level. Values are not measured against some general consensus about justice or right, but are simply defended as "my" opinion in a liberal fantasy of autonomous individuality as self-generating belief. When an opinion is valuable only because it is "mine," rhetorical manipulation is the only significant political weapon, and the only measure of right is (often) simply to be on the side of the winners.

A binary logic exercises absolute authority over the simple calculations of bit values that are the entire operation of the digital machine, engendering a world of necessity. The work of the digital is the formal manipulation of 0s and 1s according to strict rules. How do these 0s and 1s come to bear meanings beyond their empty formality? Without codes mapping bit sequences onto meaningful aspects of the human world, the computer would just shuffle 0s and 1s, performing vast calculations with no consequence. Whence the codes that coordinate (inherently) meaningless sequences of 0s and 1s with meaningful elements of the world? Not the binary logic alone, but another logic oversees the combination and sequencing of elemental bits to accomplish digital tasks, a logic that connects those elemental operations to the human world and to the desires that people bring to digital machines. This second logic, a *logic of representation,* comprises the codes that invest the empty formalism of bits with meaning. It establishes those codes by constructing and maintaining relations among bit sequences, referring them to each other and so also to the images, sounds, texts, and other forms of information that issue from (and enter into) digital devices.

The logic of representation, however, must administer these codes with a rather meager toolkit: it employs only the binary logic, building the entire edifice of the digital from bits and logic gates. A single building block, the bit, and a set of sixteen operators[9] to direct bits—it's a paltry vocabulary, but it serves to enact the codes that give the digital its significance. A code is a rule of correspondence whereby a sequence of bits refers to something outside of itself. The logic of representation makes this reference possible by organizing the binary logic so that the values of some bits can affect the values of others. At the simplest level, a logic gate, one of the sixteen in Figure 1, accepts two bits of input and generates a single bit of output, such that the output value depends on the values of the input bits. This is already a rudimentary form of reference, an about-ness, in that (the value of) the output bit indicates something *about* (the values of) the input bits. More complicated codes require many more logic gates, but the principle remains the same. For instance, a sequence of bits can act as a counter that keeps track of how many times some other operation gets performed, and such an arrangement would use logic gates to create a loop that performs the operation, subtracting one from the counter each time through the loop, finally exiting the loop only when the counter reaches zero. The bits used to make that counter would thereby be treated as a number, and that number (in this example) refers to something outside of the bits that represent it: the number of times an operation is performed. Even that simple example would require many logic gates, in part because those gates deal with bits only two at a time and not in longer sequences. The operations necessary to make a letter appear on the screen in a word-processing application when a key is pressed might involve billions of logic gates, and the logic of representation determines how those gates must be arranged to preserve the codes that map a key-press onto a sequence of bits (traveling as electrical signals from the keyboard into the machine), that process those bits according to the programs running on that machine, and then that interpret that processed bit sequence to make the appropriate letter shape show up on the screen as governed by the code that maps bit sequences onto letters. (ASCII is one such code, but there are others.)

Bits could never refer to other bits were it not for the asymmetrical difference that constitutes them, the difference between value and place. Because bits have values (0 or 1), they can represent numbers, and because each bit has a place (the nth place in sequence), it can be specified by a numerical index. By taking a value (of some bits) as designating a place (where other bits are located), the logic of representation imbues bits with an intention or aboutness. All of the bits in the digital machine are numerically indexed. Every location in RAM, on the hard drive, or in other short- or long-term storage is assigned a number that can be used to retrieve the bit values stored at that location or to assign new values to that location for later retrieval.[10] Numerical indexing undergirds the operation of the logic of representation because it makes it possible for a sequence of bits, taken to represent a number, to point to another sequence of bits stored at the index indicated by that number. At the lowest level of the digital machine, built into its hardware and prior to any particular application executing on that machine, the system already has the potential to refer to itself.

Using that referential potential, the logic of representation organizes bits into formal structures, determining where a bit sequence begins and ends and how to treat the values of that sequence so that they enact the code that gives that sequence its meaning. A sequence of bits might be associated with a pixel on the screen, storing the thirty-two-bit number that encodes the color of that pixel; the logic of representation organizes logic gates that direct the thirty-two bits in sequence to the device subsystem that lights up pixels, and that subsystem also uses logic gates to associate those bits with different electrical potentials (that flow into the display hardware) in order to make the pixel in question take on the proper hue. That sequence of thirty-two bits takes its meaning as an encoded color only because those bits are organized by the dynamic logic of representation to have the effect of determining the color of a particular pixel that appears on the screen. In the logic of representation, bits are the building blocks of formal structures, and logic gates are the articulations of those structures.

Combining many bits in articulated sequences, one could specify the color values of lots of pixels, to make a whole image appear on the screen. The complex forms constructed by the logic of representation come to represent real and imaginary aspects of the world outside of the digital system, lighting up pixels, generating sound through speakers, appearing as text or image or video, and so on, all according to codes that map bits onto other things. Notably, for all the complicated representational possibility of digital machines, all of their forms are built only from bits and using only logic gates as tools. Every operation of the machine reduces to those gates and binary values, which demonstrates that the binary logic and the logic of representation are coincident. Every bit in the machine is subject to both logics, and the two logics effectively constitute each other: the logic of representation has no vocabulary but the binary logic; it does its work only by deploying logic gates and directing bits through them. The binary logic, for its part, can neither say nor do anything without the logic of representation to set it in motion and give its bits a specific context in which they take on a significance beyond the empty formality of their values.

It's more suggestive than precise, but one might distinguish these coincident logics like this: the binary logic rules over the *values* of individual bits, determining those values through rigid calculations according to the strict rules of the sixteen logic gates. But the logic of representation operates on the *places* of bits, organizing them in sequence and arranging (through binary operations) relationships among those sequences of bits that enact the codes to which those bits correspond. The binary logic is a meaningless formalism, but the logic of representation generates meaning by responding to the actual and possible uses of the digital device. That is, inasmuch as it structures bits to forge relations among them, and inasmuch as that structuring establishes and maintains the codes that connect bit sequences to the world outside of the digital, the logic of representation answers to the human world and the many uses to which we put our digital machines. Recall Figure 1 showing that some binary logic gates have semantic correlates, such as "AND," "OR," and "NOT." These remain pure formalities until they are deployed according to the logic of representation in a context in which the objects they are comparing mean something more than just the formal difference between bit values.

Even the set of commands in a machine language for a particular CPU, even the way that the logic gates are arranged on the surface of that chip, even these generally invisible details are determined by a representational logic, constructed as such in order to make that chip useful for the sorts of things that we do with digital devices. Bitwise logic operates at the lowest level of the machine, but carries its rigid rules outward toward the interface and even beyond it, into a digital culture. The logic of representation travels the other direction, from the user's desires into the machine and even down to the digital chip, designed in accord with possibilities of representation. This is precisely why it is a logic of *representation*: the organization of the binary logic is driven by the representational demands that we bring to digital machines. The design of the whole digital system, the organizing principles that govern software and hardware, is entirely determined by the representational uses to which the machine must be put, the things we want to do with it.

Face as Icon

The flattening of meaning in the digital appoints the *icon* as the model of digital significance. Evacuated of indeterminacy, everything in the digital exhibits the rigid, conventional logic of the icon. We are familiar with their spread on desktops and in application toolbars, but icons populate the entire interface: a discrete (usually visual) unit, responding predictably to defined inputs, and typically representing as proxy some other digital object or operation. According to this logic, menu items are icons, as are Photoshop filters, but so are, for instance, the expressions on the faces of digitally animated characters in video games. Each character's face is defined by a collection of points in a shallow three-dimensional prism, and a given expression—joyous, quizzical, distraught—is achieved through a formulaic translation of those points within that space. Though this algorithmic representation of emotion risks reducing expression to homogeneous, *iconic* facial dynamics, it feels even more troubling that the expressions themselves are generated from categories. When a real person pouts or sneers, whatever is on that person's face is not confined to a generic version of petulance or contempt, but bears all the contingent complexity of a life and its context, such that every facial expression is necessarily a complex mixture, a whole history of affective dynamics. An expression of fear on a person's face is never pure, never finally iconic, except in certain works of art or other idealizing media. With a computer character, on the other hand, the configuration of the face is nothing more than the assertion of the current emotional state: the character is now feeling fear, now ecstasy, now consternation, and we are trained to read those faces and the stories they tell according to that categorical logic. That is, the vocabulary of icons that defines the range of possibilities for a character's facial expression restricts the narrative possibilities for digital storytelling, such that stories are told as a sequence of icons: each character at any given moment is *either* neutral, or happy, or irritated, or astonished, or horny, and so on. Computer characters rarely feel real, regardless of how many polygons they comprise on the screen, because they behave within the constraints of this iconic logic, representing preconceived states of mind through calculable, legible facial dynamics.

Even when motion-capture ("mocap") has provided the underlying data of facial expression, the face tends to convey predigested meaning. The mocap system has already designated certain points that are considered the distinctive bearers of meaningful gesture (facial or bodily), which are the points tracked by the mocap system. Then the data are automatically reduced to make processing more straightforward, which means that not only are many complementary movements of the face or body not being captured to begin with (though these are the subtle ones), but then the remaining subtlety is filtered out of the mocap data in favor of gestures that can be formalized or formulated meaningfully. Finally, the animators will edit out (or in) certain gestures in order to ensure that the character's meaning will be legible as intended. If a character is supposed to appear quizzical but the captured mocap data fail to convey this affect unambiguously, then it is easy enough to borrow other data or simply write new data to signal puzzlement and curiosity by that character. (And there has been plenty of research that matches generic facial and bodily gestures, in the form of relations among points of the face, to categorized emotions or affects. These data can be used to make a generic face express a generic affect, which is pretty much how digital expression operates.) The reduction of expression to formula is especially evident in the animation of speech; whereas digital animation must choose a finite number of points to set in motion on a character's face, "the pronunciation of even the smallest word," according to Ferdinand de Saussure, "represents an infinite number of muscular movements that could be identified and put into graphic form only with great difficulty" (15).

A sequence of bits might stand for a visual pattern (for example, a checkerboard pattern), but that "standing for" does not inhere in the bit sequence; it must be actively maintained by an algorithm that takes the bit sequence and uses it to instantiate that pattern visually (or virtually). A bit sequence might represent an image, say a person's face, but it does so only by virtue of an algorithm that encapsulates the relevant encoding and effectively decodes it to produce that image on a screen or printout. A bit sequence might represent a command or a whole program, but that representation requires a system designed to take the bit sequence and execute the represented commands by directing those bits to the appropriate parts of the hardware mechanism. The logic of representation, ultimately serving the representational possibilities of the machine, determines which codes are implemented and how. That logic, built from the binary logic and limited to its possibilities, governs the creation of codes and brings them into existence by embedding them in the operation of the machine.

These two coincident logics, a formal binary logic and a logic of representational form, determine the ontology of everything in the digital, for it is all made of bits articulated by logic gates. Bit sequences map isomorphically onto elements of the interface such that the whole of the digital demonstrates an ontology borrowed from the bit. From the most complicated to the most elemental digital structures, all is discrete and all is assembled from discrete parts, themselves made of discrete parts, and so on, until the analysis comes to rest at the bottom line of the discrete bit. Codes determine the parts from bits, and codes determine the whole from parts. Images are made of pixels; word processors format text into lines, paragraphs, sentences, words, sections, pages, documents, chapters, fonts, font sizes, languages, and so on. A video game allows the player to switch among four different weapons. Moving the mouse wheel clockwise zooms the image or scrolls the document or increases the numeric index under the mouse pointer. The database associates nine fields with each record: three text fields (street address, city, and state), four numeric fields (zip code and birth year, month, and day), a binary field (male or female—it's a twentieth-century database), and a field containing each record's unique identifier. The email window has three columns, and each email is split into a header and a body. An internet protocol (IPv4) address has four eight-bit indexes, each of which is a number between 0 and 255, though this is just a more legible way of representing a thirty-two-bit number. Photoshop divides images into color channels, layers, masks, selected and unselected pixels, luminance, and chrominance. These articulations or mappings of the digital each responds to a logic, a set of principles and practices that determine how a domain is filled out, what things and what operations are possible in a particular digital space. The digital interface to a toaster has a logic, *toaster logic,* typically articulated by the buttons and dials that control its various states and degrees. Not every logic need be deterministic or rigid, though the digital logic is both.

The digital thus emerges at the intersection of these two logics, one an empty formalism and one that arranges that formalism into structures that reflect the outside world. In the digital, a person is data about a person, a narrative, visual, sonic, or statistical representation of a person, a detailed list of characteristics of a person, the stylistic categories that include a person, but all of that is always only formal structure, a set of relations among otherwise empty formalities. Standing at the antipode of formalism is contingency, which can appear only as a disruption of form, an indeterminacy as yet unspecified, that which the digital cannot capture. The logic of representation erects forms from the empty formality of bits, and those forms mirror the forms of real and imaginary things and events out in the world, things and events that also have a form but that, thanks to their immersion in contingency, give substance to that form.

Reducing being to the simple assertion of a posit, *form* is being, but it is a being stripped of becoming, stabilized in its self-identity, thrust forward as a definite thing. The digital, master of form, can capture almost everything, almost the entirety of being, denied only being's elusive reality, the contingency that gives being its meshy substance, exceeding formal definition. The digital is an Eiffel Tower, all girders and rivets, all steel, hard edges, and monochrome, and the world lights up in reflection, glittering Eiffel Towers everywhere, skeletons so magnificently illumined as to scorn their absent flesh. The logic of representation takes worldly forms as its model, constructing sequences of bits, twisted and bent to match the complexity and the structure of those real forms that surround us. This is the formal relation that is also the content of the digital, and the digital speaks to the world by an approximate equivalence of its structures with those that articulate the world.

Form admits no ambiguity: it is one way or another way but does not hesitate between ways. It is *there,* or it is the sort of thing that could be *there,* unequivocally, definitively. Measurable and discrete, self-identical and absolute, digital forms, manifest in the structural dynamics of sequences of bits in relation, can correspond to worldly forms or can fail to do so, but form is always what is correct (or not), what corresponds (or does not). "What is neither true nor false is reality," says Jacques Derrida (197), and it is that untestable, unmeasurable, uncapturable reality that always eludes the digital and its formality. The digital carves its forms out of the actual but does not encounter there the generative principle, the contingency that races ahead, leaving form in its wake, following along like contingency's shadow.

Jacques Lacan worked for much of his career with a tripartite schema to describe the human world: the *Real*, the *Imaginary*, and the *Symbolic*. And these terms map closely onto the present discussion of the digital. Lacan's schema would assimilate the digital to the *Symbolic*, inasmuch as the digital operates internally by the manipulation of symbols, 0 and 1, or rather the asignifying bit values that we call 0 and 1, which stand for nothing but await the assignment of meaning through (symbolic) representation. Form is symbol, at least when it is related to other forms. The *Imaginary* domain connects forms to each other, establishing identities and similarities, and so giving the symbolic a value that exceeds its pure formality. Thus it is through the Imaginary that the computer can say something about the actual world and can receive instruction from and offer instruction to people. It is an effect of the imagination that we can relate a simulacrum to the simulated, or more generally that we treat the machine as though it were a subject. Lacan's elusive idea of the *Real* corresponds well to the notion of contingency described in this book. It irrupts without pattern, without assurance, as the uncapturable, the uncontainable. It exists, but only under the condition of its own erasure. The Real is implicit in the dynamics of imaginary and symbolic, which could not exist without it, but it refuses to submit to any positive account, escaping symbolization, eluding imagination. The Real is the complement of the digital, that not-quite-something that makes everything what it is but hides from all scrutiny, marking the digital's inevitable limit.

At the intersection of the two logics, binary and representational, the digital machine reveals its general nature as *a machine of representational possibility*. The logic of representation determines how to encode an input as a sequence of bits and how that encoded input can be manipulated using binary logic to generate the appropriate output. Software generalizes this concept, providing a standing sequence of logical calculations that accepts an input, breaks it down into sequenced bits, and then runs those bits through those calculations to yield an output. Software, like the digital machines that host it, thus functions as a space of possibility, an environment in which a particular restricted set of possibilities is made available.

Between the interface and the individualized bits, digital objects and digital processes imply an intermediate layered space of successively more complex organization. A complex edifice of digital structure arises, as codes govern bits but are themselves governed by other codes, and so on. To make a menu or a piece of software or a clickable icon, one does not just gather bits into grouped sequences. Rather, one builds from bits to complex structures, starting with (relatively) short sequences of bits that make simple structures, then building more complicated structures from those simple ones. A sequence of eight bits makes a *byte* (or at least that is the convention employed in almost all modern computer systems), and bytes can be sequenced to hold simple data of different types, which are often represented as labeled variables in a computer program. Variables or other simple structures made of bit sequences can be grouped together to make more complex structures, which can themselves be grouped to make still more complex structures of data, and so on. The examples offered thus far, like colors, machine language, and ASCII, show how representations can be directly encoded as sequences of bits, but much digital structure occupies intermediate layers, organizations of bits that make higher organization possible. The data structure associated with an icon visible on a desktop might, for example, include a collection of different data structures, one structure that stores the image of the icon, a variable that stores its current location on the desktop, another that stores its text label, another that stores a pointer to the algorithm to be executed when it is double-clicked, another that stores what type of entity it represents (a file, a program, an action, a passive image, etc.), and more. A window open on the desktop likely has even more numerous and complex data structures associated with it. This dizzying mereology derives from the fundamental ontology of the digital: one builds objects and processes in layered steps, from discrete bits, to discrete groups of bits, to discrete groups of groups, and so on, until one arrives at a form that mirrors the form of the object being digitally described. Layers of structure and their interrelations are the concrete expression of the logic of representation as it encounters the bitwise, binary logic.

Not only data but also processes are structured in layers, as elemental logic gates are grouped to carry out the more involved commands of machine language, which are in turn grouped to carry out the still more complex commands of source code, which are themselves organized to carry out commands chosen by the user in the interface. To create a piece of software, one must break down the (desired) capacities of the software into smaller tasks, and those into still smaller tasks, to arrive at a set of algorithms or subroutines that collectively carry out all of the desired behavior of the program. Those subroutines must be further incrementalized until the entire program has been described in the commands of the programming language one is using. And those commands are then broken down automatically, without the programmer's direct intervention, into individual commands of machine language, which yields an executable piece of software. Crucially, each step of breaking down a task into smaller subtasks must grapple with the constraint that the process will ultimately issue in sequences of simple commands. The programmer cannot haphazardly divide the operation of the program at the outset into subtasks, but chooses the organization of subtasks with respect to the representational logic expressed in the coding language, which itself must ultimately respect the underlying binary logic of the machine.

Given that an algorithm might muster billions or trillions of sequenced logic gates and billions or trillions of bits to pass through those gates, the possibilities on offer for creating and manipulating form are mind-bogglingly complicated and potentially detailed. Formal codes are most easily grasped through static examples, such as desktop icons or color values, but an algorithm in its dynamic process is also a set of formal relations. All the parts of an algorithm are entangled with each other through the representational logic that relates place to value, which can make an overview impossibly complicated even if a step-by-step analysis of the algorithm can be performed by a patient human. The point is that the forms by which the digital device makes meaning in relation to the actual are not confined to structures of sequenced bits standing still, but include the prismatic and fractalized network of all those simple gates and all those binary bits interrelated in a structured knot of complexity.

Computer Chess

Is it possible to become a master of chess by playing against a computer? One could surely hone one's game competing against a chess program, but as Hubert Dreyfus points out in his 2009 *On the Internet*, the closer one gets to mastery, the more the software's limits would also be the limits of your learning. For to play against a computer is to play against a strategy that (ultimately) does not change. Even software designed to exercise a variety of strategies would be playing according to an algorithm that has already chosen criteria that determine what it considers the *best* or *right* moves. Those criteria might include a kind of metachoice among different ways of making that choice, such that the criteria for choosing a move sometimes favor one strategic decision, say aggression, and sometimes another, say defense or feint. Or something far more nuanced may be programmatically implemented, approaching the sorts of criteria that human chess masters use. With a human being, it is always possible that they will see a strategy they have never before seen, develop criteria that did not previously exist. A computer can deploy only those strategies, execute only those decision criteria that it has already been programmed to execute, even if those criteria determine a move that is, objectively, unexpected, in the sense that chess experts have not previously seen it or regarded it as a wise choice.

But one of the glories of chess is that there are enough possibilities of play, and the choices are sufficiently numerous, that humans can actually see something they have never seen before, can make a choice to take a path commended specifically by none of their previously employed criteria but that still might be a good path. Human beings can learn, whereas the computer is fundamentally rule-bound, confined to its program.

But software can also be programmed to do certain kinds of learning. Chess programs are routinely given the ability to learn from mistakes, keeping track of suboptimal choices that led to undesirable outcomes. Software can be developed to recognize something like context, at least in the delimited domain of chess pieces and moves, and can learn to decrease the weight of previously failed choices when confronted with a context judged similar to the one in which those failed choices were made.

But contextual factors must be specified in advance for the computer; it must be told which things about the chess situation to measure and what formula to use when comparing the measurement of a current situation to measurements of previous situations in order to gauge similarity. Such a comparison indicates similarity or difference and can then be used to decide how to weight various strategies, discounting strategies that in prior similar situations led to bad outcomes. But at some point the formula is invariant. The decision about which criteria to use to choose the best move, or which criteria to use to decide which criteria to use to choose the best move, or which factors to use and how to measure them and compare them in order to choose which criteria to use and how to weight them—all such decisions are given by predetermined rules that do not change. The buck always stops; the software is finally unable to invent something new, to see something a new way.

This is partly because the software can't really see anything at all. It can only measure. Thus, the computer falters when asked to account for why it makes the move it chooses. It can of course be programmed to analyze its own criteria and weights and then narrativize the corresponding numbers to present an *explanation* of its decision. But the decision is first of all driven by the statistical weighting of the underlying numbers, whereas the narrativized reasoning is after the fact, a result but not a cause. And even if one could design a chess algorithm that figures out its next move by manipulating some sort of narrative of the situation, it too would be bound by prior numerical rules that determine how the narrative is turned into numerical weights and calculated to yield a statistically optimal next move, according to the preestablished calculation model and its criteria.

Note that this discussion refers to software decisions that are not readily resolvable by *brute force*, by looking some number of moves into the future. Many moves can be eliminated as not worth considering when it is calculated that a certain move results in a situation easily dominated by the opponent. But when that look-ahead technique encounters situations that have no clear best move, then other criteria, such as models and weights, must be applied in order to decide which move to make next. Indeed, the look-ahead method is simply the first and primary set of criteria in any chess algorithm, as for instance a move that allows the opponent to achieve a checkmate on the next move is clearly a relatively poor choice and can be given a nearly zero weight. (If no move exists that avoids a checkmate for one's opponent, then some such poor move must be made, or the software can be designed to resign at that point, but in any case, a move that leads to checkmate still gets assigned a nonzero weight.)

All of this is to say that playing against a computer is always playing against some already decided rule. Unlike a human being, a computer cannot be truly spontaneous in its creativity.

M. Beatrice Fazi argues that this is the wrong way to look at it. She proposes that, every time the algorithm executes, even though it is in some sense bound by preestablished and invariant rules, it is venturing into the indeterminate, generating something new. This is because, Fazi explains, there is no algorithm that can predict what a given algorithm will do. In particular, there is no algorithm that can reliably calculate whether a miscellaneous algorithm will eventually terminate or will continue without ever reaching a terminal condition. Therefore algorithms generally have something undetermined about them, an ontological reservoir of indeterminacy. And the execution of the algorithm consumes this indeterminacy as it proceeds, determining each next state of the machine, including the terminal state of the algorithm, which in many cases is the output, as something that was not foregone prior to its production. She does not argue that the rules might be broken, that an algorithm has unlimited possibility or might do anything, but rather that there is a certain kind of contingency in the algorithm, an openness, yet to be determined each time the algorithm executes.

Fazi's is a powerful and expert argument: the rules in the digital may be absolute and unyielding, or determinative, but the process and its outcome retain a measure of indeterminacy. The question is whether this formal contingency is satisfying, whether it offers the creative foment, the sustenance of novelty identified with actual contingency, or whether the absolute contingency that respects no rule is in fact what we should value. Computational contingency, if it exists, is the claim that whether the machine outputs 0 or 1 is fundamentally undecided by the universe until the program actually runs and produces an output. Actual contingency is the claim that the universe itself never knows which of its infinity of aspects might enter into a given situation. Are these meaningfully equivalent contingencies?

Notwithstanding the insistent characterization of the digital as positivist, the analysis in this chapter reveals a tension between this ascription and traditional understandings of positivism. For one thing, historically, and especially in the philosophy of science, positivism has been closely allied with empiricism, whereas the digital, as opposed to the actual, largely operates in an abstract space askew of the empirical. The digital nevertheless merits the label *positivism* because of its discrete, autonomous elements that, like basic empirical facts in relation to theories of physics or chemistry, function as foundational givens within digital systems. Bits and their values constitute a digital bottom line, immune to question, indivisible, unambiguous in their formal significance, and determinative of all more highly organized phenomena in the digital system. But even if *positivism* is pushed back into its etymology and separated from its historical association with empiricism, the digital's formalism calls its positivism into question. A posit is by definition something that is fundamentally, unequivocally, metaphysically *there*; it claims for itself a plenary reality; its being relies on nothing else, but asserts *itself* first of all. Yet bits, as described throughout this chapter, are in many ways nothing whatsoever; they are stripped of meaning, merely formal, and mostly abstract. Positivism made digital has been evacuated of one of its central pillars, the there-ness that is part of any posit. This formal positivism, an insistence on the bit as a bottom line, retains the positivist fortitude of indivisibility, autonomy, and simple undeniable assertion, but what gets asserted is practically nothing at all, only the formal difference of value between 0 and 1. The digital exhibits the form of positivism, a performance of positivism, without positing anything one could bang one's shin on. It shouts *0* and *1* into the void, a hollowed-out, anemic positivism that requires prostheses to muster any real force.

Nor is it only positivism that is hollowed out by the digital. Rationalism, too, becomes merely performative, abandoning in the digital its weightiest principle, the spontaneity of reason. If there is a sufficient reason for everything, if everything in the actual happens for a reason, this principle still leaves room for the invention of new reason, where reason might engage something genuinely unknown according to the ordered world's beautiful and transcendent complexity. But, in the digital, reason is a mechanistic guarantee, appealing no farther than the truth tables depicted above. Everything in the digital has its reason, but it is always the prescribed reason of the digital game, the reason of a world in which reasons have already been chosen and can never be spontaneously invented. A desire fundamentally unanticipated by the software designers is likely a desire unfulfillable by that software. Updated software or a new algorithm opens new possibilities, but that reason must first be traced out in the machine and does not arise spontaneously in its users.

Digital instrumentalism is just as true but also just as shallow as its ideological brethren. Certainly the digital gets things done, in short order, in large quantities, and around the world; it is in many ways the ideal instrument. But its hermetic enclosure, sealed in its own world, ensures that its operations remain impotent, confined to its virtual space, unless and until it is connected by means of prostheses to the actual world beyond its borders. Any digital action, always a sequence of calculations, takes place in its abstract realm, such that its consequences must be only formal results, waiting to be acted on or responded to by agents outside the digital world.

This analysis depicts a barren digital, with an eroded contingency and a central ideology of values stripped down to vacant and superficial form without content. Doesn't this fly in the face of our nearly constant experiences of digital technologies, which seem today to host the greater part of what is new, exciting, and significant in our lives? How can the digital, the backdrop and substrate of much human endeavor and the modern medium of desires grand and casual, turn out to be so entirely empty and without substance? These words—*empty, vacant, superficial, hollow*—suggest an extreme position, with the aim of revealing a digital paucity that we usually and carefully hide from ourselves. The digital is truly hollowed out, but even without the assertion of substance at the core of positivism, even with no materials beyond the rarefied purity of form, a great deal can be represented. And representations, as evinced by the astounding success of digital technologies, are extremely powerful tools.

The digital produces only representations, but representations convey more than the formal elements they comprise. The digital is thereby able to carry meaning that it does not store or act on, meaning that cannot be found in any of its bits or formal structures. Text, a primary form of digital input and output, offers a simple illustration that can be easily extended to most of the digital's other forms of representation. The ASCII code mentioned above encodes a limited but highly useful set of text glyphs, such that an algorithm can input text, store it, search it, display it, alter it, compare it, and more, all without any (simulated) awareness of the text's *meaning*. A string of characters might constitute a message with profound significance to a human reader, but the digital machine need have no capacity to understand that message to be able to manipulate the character string in countless ways. The possible manipulations are limited to those facets of a representation that submit to positivist encoding. It turns out to be challenging, for example, to instruct the machine to take a text message and make it slightly more emphatic or a bit less cruel, because those characteristics are difficult (though not necessarily impossible) to capture in general as positive (measurable) traits. But, as emphasized earlier in this chapter, much of the world (some would say all) is amenable to positivist description, such that, for all the digital's hollow and insubstantial formality, it remains reliably effective, capable of accomplishing just about any definitive (representational) goal, and thus giving rise to all the nuanced and complex meanings that representations can carry.

Rendered as image (or sound, or text, etc.), representations are then exposed to human practices of vision, audition, reading, and more. That is, the prostheses that connect the digital machine to the human world beyond the digital respond to the ruliness of the digital, respecting its formal limitations, but they also move beyond those rules into a phenomenal and material domain. Looking at an image on a computer's monitor or listening to the music from a compact disc, our creative apperception can see and hear more than (and other than) the data that constitute those perceptible representations. Today's digital interfaces retain much of the superficiality of the underlying digital formalism, but in principle (and there are signs that the industry is moving in this direction), the digital machine will build out of its forms vast three-dimensional worlds and possibly even means of inhabiting them. How will the diminished contingency of such manifold simulated spaces make itself felt?

❨ 5 ❩

From Bits to the Interface

Two worlds, actual and digital, each with its own ontology; the difference in tone between the two ontologies is dramatic. The actual seethes, slips, and oozes, denying any clarity but the momentary and ephemeral. The digital offers nothing but clarity, individuated bits, grouped together to form a layer of basic structures that are themselves grouped to make further structures, and so on up through the layers, to form more complex, multibit digital objects, each of which retains the absolute discreteness, clarity, and definition of the bits it is built from. The actual rests on no foundation, no elemental unit, and even the image of a mesh really only points the way, a metaphor of materiality to help depict what no picture can show. The actual can be equally understood as an encounter of grand forces, generating unending subtle difference in their interactions, or as tiny even infinitesimal differences that combine and resonate to form semistable boundaries, conditions of enduring, patterned dynamics. The digital, by contrast, is built bit by bit, layer upon layer, and its stunning numeracy generates a complexity that belies the total simplicity of any of its layers or objects when viewed up close.

This contrast grows still more stark as we examine the layered structure of the digital and the ways in which it is determined by its bitwise foundation. Each layer carries the ontology of the bit to the next layer up, such that digital objects and digital processes exhibit the same absolute definition and discreteness as do the bits from which they are built. Because bits are discrete, self-identical, and individuated, so are bytes likewise discrete, self-identical, and individuated, up through the layers, so that menus and windows and desktop backgrounds also are discrete objects, with precise values and perfectly defined boundaries. As we have seen, the bit arises at the confluence of two differences, one of value and one of place, and this difference of differences percolates through the layers of the digital, stamping each layer with a split ontology: each layer of digital structure gathers a group of structures from the layer below into a more complex structure, but the bits that are thereby placed in that complex group of groups also have values that are determined independently of the structure itself. Each structure thus serves as a *part* for layers above it, whereas the values of the bits included in that part determine its *quality*.

A brief example: A digital hamster has four legs, each one a part represented by some structure of sequenced bits. But the values of those sequenced bits would determine the qualities of the hamster legs, such as what color or how long is each one. A more extended example: A software-illustration application might include a tool that allows the user to draw a quadrilateral shape in the application window. To implement such a tool in software, the program code includes a data type that serves as a general description of a quadrilateral. That general description is specified in the source code as a structure without specific values, a description of groups of bits (or groups of groups of bits, etc.) that will be given specific values when the program is run but that remain unspecified in the source code as written by the programmer. The generic quadrilateral in the source code thus serves as a template or abstract description. It may incorporate four data fields, one for each side of the quadrilateral shape. Each of those sides is a *part* of the quadrilateral, and each also has its own parts, its endpoints. But a side may also have other data associated with it, including information about its thickness or texture or color, and those data are given, for each specific quadrilateral drawn by the program's user, by the values of the bits associated with each side (part) of the quadrilateral. As modeled in the source code, the parts exist independently of their values; the structures that constitute the abstract parts are designated as sequences of bits with values to be determined when the program runs.

As discussed in the previous chapter, this kind of structuring implies layers, where the individual parts are on one layer and the whole to which those parts contribute are on the next higher layer of organization. The sides exist on one layer, and the quadrilateral on the next layer up. The quadrilateral too might be a part of some larger collection of drawn objects on a still higher layer, perhaps one pane of a pictorial representation of a window with multiple panes. Each layer is made of qualified parts, but in contrast to the ontology of the actual, those parts and their qualities bear no inherent relationship to each other, nor to the higher layer that they constitute in aggregate. In effect, a bit's place in a sequence assigns it to a part, and that bit's value contributes to the part's quality, but there is no necessary connection between the placement of a bit and its value. Parts are essentially shaped by the logic of representation that imposes structure from without, whereas qualities (values) are given by the logic of the binary operating at the lowest level of digital calculation.

Place and value, part and quality: rather than an immanent co-ordinated evolution that determines these things in an inherent relationship to each other, as in the actual, the digital treats them as independent entities. A real hamster's leg does not develop without relation to the rest of the hamster, and the leg's various qualities, such as size, strength, color, and furriness are not independent accidents accruing to the leg, but are all interimbricated and dependent on the whole hamster. In the digital, a part and its quality can be algorithmically interrelated to simulate a degree of imbrication, but any such relation must be deliberately inscribed in the algorithm and will extend no further than this deliberate inscription. This positivist autonomy might also be described as a disconnection between *form* as the structure that makes a part and *content* as the values that determine the part's quality. In the digital, form and content do not develop in relation to each other, but must be explicitly connected if they are to have any relationship at all. This is not to contradict the previous chapter's assertion that the digital is all form and no content; even the values of bits are only formally distinct, and even the difference here articulated between content (quality) and form (structure) is itself a formal difference, a heuristic and relative distinction.

Layering supports the logic of representation by providing access to structures of articulated bits without having to treat each individual bit within the structure. Consider a digital layer just above the plane of bits. On its out-facing side, it gathers groups of eight bits from the plane below it into a sequence, organizing those bits into bytes, which allow a ready treatment of eight bits as a provisional unit. Defined as a group of eight sequenced bits, the byte is abstracted from the values of those bits and can be treated in digital code without those values being specified. This abstraction is a basic method of digital operation: in program code, a byte can be designated, manipulated, copied, multiplied, and so on without making any assumptions as to the values of the bits within it. Facing the lowest layer of bits, this byte-wise layer also includes a value for each of the eight bits, filling in the eight-bit byte inscribed in outline on the layer's outward-facing side with one of 256 possible values. (There are 256 different combinations of 0s and 1s that can be made using eight bits in sequence, or 2^8.) The formal outline of the byte is ontologically distinct from the values of the bits it gathers within it.

This heuristic distinction between two sides of the same layer, between a part and its qualities, or between a structure of bits and the values of those bits, often also tracks a (rough) division of labor and time. Whereas the programmer specifies the form of a generic object, defining the *parts* in relation that it comprises, the user usually determines the *qualities* of those parts, taking actions that assign bit values to the sequences of bits that constitute those parts. This was illustrated above with the quadrilateral, the general outline of which is given in the program code, but the details of which are decided (later) by the user, who draws the particular shape and chooses the qualities of its parts. This division of labor is not a hard rule: there are plenty of objects determined in the source code by the programmer even down to their qualitative details, and there are also opportunities in software for the user to construct new forms. But it is nevertheless characteristic of software that it offers a prescribed general template (or many of them) to which the user adds specific content. Programmers and users both input information into the machine to delimit possibilities and generate outcomes, but they are at least approximately distinguished by the level at which they operate, as programmers tend to work with generalities and users more often with specifications. This fills out the sense in which the computer is a machine of representational possibility, fixing a form in software and offering the user any number of possible options for customizing it. Word-processing, tax/accounting software, CAD (computer-aided design) and CAM (computer-aided manufacturing) applications, three-dimensional modeling, even browsing the web—in each case, the user makes specific choices against a fixed general background, filling in a template with qualities that, prior to her choices, are only possibilities.

The general pattern of digital operation begins with abstract representations that accrue specific qualities through independent processes. There is thus in the digital a *theory* of each object, a fixed set of guidelines that determine the object's general form and the range of possible values that can fulfill that form, giving it specific qualities. Programming is the act of inventing theories of representation, defining a world of things and their degrees of freedom or variability. In the digital, there is always a rule that determines what counts as a given object, along with the degree of variation and the nature of the variation allowed in specifying that object. There is a general conception that decides what counts as a quadrilateral in a drawing program, or what counts as a paragraph in a word processor, or what counts as a tax-deductible expenditure in an accounting application, and any actual instance of those things must fit the established rule.

Inasmuch as the codes that endow bit sequences with meaning establish correspondences between individuated bit sequences and external referents, the logic of representation performs its operation piecemeal. Beginning with the bit, every structure, every part, every quality, every value, every process, every command, everything in the digital is discrete and independent, bearing no inherent relation to any other thing. A digital object (or process) is built from parts, themselves objects, that are conjoined to form the larger object, much like puzzle pieces make a puzzle. But the values of those parts, the values of the bits that are grouped to make those parts, bear no necessary relation to the process that organizes those bits into a group, as though each puzzle piece had an image on its surface formed independently of how the pieces all fit together. Any relation in the digital, any connection between bits or higher-level objects must be explicitly created in coded instructions. This a-relationality or positivist independence is closely connected to the inherent meaninglessness of a bit. To have a relation to another bit would give a bit at least a modicum of significance, some meaning that exceeds its empty formality, whereas its autonomy ensures its senselessness and renders it available for any significance imposed from without.[1] Likewise, parts and the values that determine their qualities are related only when program code contains explicit instructions that maintain that relation. Everything in the digital is independent, and this resolute atomism, this extreme positivism, is mitigated only to the extent that deliberately instituted rules in the form of coded instructions establish and maintain connections among bits, or between a part and its quality, or among an object's various elements, or among different objects. The requirement of a rule for every relation in the digital, the totalized determination of every part of a digital thing, leaves no room for accidental relations, happenstance, or mutual determination. Which is another way of advancing the central thesis of this book: the digital abjects contingency.

Compare again this piecemeal digital structuring, shared by all things digital, and the coherence of ordinary objects we encounter in the actual world around us. In the actual, things are constituted and sustained in relation and by virtue of relation. The parts of a thing are not just an unrelated collection of fungible stuff. A thing and its parts all relate to each other in a tangle of dependencies, tensions, and resonances, a mesh, such that one cannot alter a quality or a part without the whole thing changing in response. Actual things enjoy a relative stability born of their generation through the struggle of indefinitely many forces that hold them together in an evolved equilibrium of interdependencies. Contingency threatens to sever some of these dependencies, or to impose a new order that rearranges the standing relations into a new stability, but the thing, in its evolved equilibrium, asserts an inertia that can push contingency to the margins. By contrast, the parts of a digital thing are sparsely interconnected by only as much relation as has been explicitly coded into that digital thing. They can be easily swapped out for other parts or their qualities drastically altered without dissolving the merely formal integrity of the whole. Indeed, the possibility of swapping out parts is heralded as one of the core strengths of the digital, a digital *modularity* that enables the use of software plug-ins, reusable code, and team-based and task-based software engineering, and that guides much hardware design as well. As everything in the digital decomposes into intrinsically unrelated atoms, even space and time submit to this assault on their integrity. The atomization of space and time manifests the prodigious power of the digital, as it regularizes, and so tames, the contingent interrelations that safeguard the singularity of the actual. What sorcery but digital operation allows an arbitrary control over space and time, the foundation of digital simulation? As representations made of unadulterated abstraction—elemental lists of numbers and means of modifying them—space and time do not exert in the digital the sorts of inexorable constraints and directed possibilities that they do in the actual. Solar systems and molecules appear side by side on the screen, thousands of simulated years can be compressed into a second, or a complex process can be slowed down, stopped, reversed, or resequenced for observation and analysis. Simulated objects, untethered from a materiality that would give them particular characteristics, need not behave like the objects they appear to be: a planet can be weightless, an autarch can oversee millennia of continuous history, a refrigerator can transform smoothly into a set of keys.

Which suggests another pairing (along with form and content, part and quality, place and value) that demonstrates the striking and radical consequence of digital positivism: appearance and behavior. Unlike in the real world, there is nothing inherent in the digital that ensures that an object's behavior will correspond meaningfully with its appearance. A digital cello, or what appears on your computer screen as a representation of a cello, could make the sound of machine guns, or crying babies, or a clarinet. (Or it could make no sound at all, if there is no code included to generate its sounds.) The sounds produced by the digital cello, if any, depend entirely on what sound-generation software the programmer writes (or borrows as a code module from elsewhere) and those sounds are in principle unrelated to the appearance of the cello on the computer monitor, which is determined by image-generating software independently of the sound-generating software that makes the sound. That same digital cello could have strings that appear to be made of crocodile teeth, but this need have no effect on the sound produced, because the algorithm that determines how the cello appears is intrinsically unrelated to the algorithm that determines what sounds it produces. Its body can be made of (simulated) lead or cotton balls, can be as big as a mountain or as small as a molecule, but unless the code is specifically ordered to reflect such changes in the cello, those changes will make no difference to the way it sounds.

Icon II

A different terminology, focused more squarely on the interface, could locate *the icon* as the universal model of a digital object. Everything in the digital operates iconically. For one thing, the digital can only represent but never present; with discrete, finite numbers as the language of action and object in the digital, every action and every object is necessarily reproducible, generic, a type rather than a singularity. A representation in the interface stands in place of a phantom object to which it refers. In fact, there is no object, only representations that successfully imply the object, making it real enough for us users. We see and manipulate an image that is never quite the object itself. We are given handles, buttons, tools, but we don't strictly speaking encounter the object, which is projected as a representation into the screen (or speakers, or printer, etc.) according to substantive rules that serve as an interpretation (theory) without any interpreted (thing). An icon is precisely such a preinterpreted representation, a symbol that mediates access to its object, a sign that signifies by virtue of an established code.

Everything seems to work this way on the digital machine. Icons always represent a conventional significance. That is, it has already been determined that the icon represents a certain file, and the symbolic significance of the icon has little (or no) room to wiggle, to slide in that way that makes meaning possible. Far from relying on a fixity of relations, the very possibility of meaning depends on a sense of the unfixed, the spontaneously creative, the chance that a new relation will accrue. Meaning requires contingency.

Computer narratives too are iconic. Digital facial expression and body language are iconic. In a digital game, dialog tends to be iconic, in part because the dialog is always mostly functional. That is, dialog needs to advance the story or fill in essential elements of the narrative, which is mostly accomplished through play under the control of the player. Thus cut-scenes tend to condense a kind of reduced narrative line, confining dialog just to essential delivery of meaning. Of course, narrative is iconic in many other media too, including film, which is an industry that often (though not always) benefits from making meaning unequivocal, so that the relevant moral calculus is simple and legible. Mainstream filmgoers expect things to work out according to established formulas, and mainstream filmmakers wish to leave this formula unambiguous, to satisfy the broadest possible audience.

This makes narratives specifically designed for the computer frequently unsatisfying or impoverished. A motivated player/viewer may well feel satisfied by the iconic message; relatively unambiguous meaning or meaning fixed by the formula implicit in an icon can still carry powerful messages, at least to those people who understand the iconic gestures. But to the extent that meaning involves creation, to the extent that meaning relies on contingency, icons eliminate the margins at which that contingency would operate, removing the uncertainty of meaning, such that we derive the meanings of digital narratives according to established formulas, just as we know exactly how to manipulate a file system through the icons presented on the desktop. As Jim Campbell notes: "Icons are designed to be precise and accurate and discrete—on or off. They are designed to present a closed set of possibilities. They are not capable of subtlety, ambiguity or question. An interface of choice and control makes sense for a word processor, an information retrieval system or a game, but not as a metaphor for interactivity or dialogue" (133).

One might wonder whether iconic narrative production really sacrifices any possibility of making meaning. In a medieval morality play, for example, are the iconic characters (named Virtue, Justice, Beauty, etc.) the sources of meaning according to formulas, or does the play work by virtue of a subtle tension between these icons and their actual roles and behavior on stage? On the computer, is the narrative value generated by iconic meanings? Or rather does it emerge in an uncanny tension between an icon's generic meaning and its aesthetic excess, signifying beyond the conventional meanings associated with that icon?

We might explore the proposition that the icon has itself become iconic, a primary symbol of our digital age. So much film and television and so much politics and media communicate through images, gestures, phrases, whole styles chosen because all too familiar. Much of culture now bases itself on the meme. If a remarkable object is one that, at least at its margins, feathers into a subtle but arresting difference, then culture today has contracted, pulled back that productive difference, as though ashamed or even chastened at the prospect of making noise. No doubt corporatism contributes significantly to the dearth of original thought that manufactures today's digital culture. It incorporates culture at large, with its peculiarly American liberal emphasis on the inherent (but formal) value on the individual who fits into a rigid hierarchy of roles, each of which is assigned a circumscribed domain of thinking.

Then again, culture has always relied on convention, and is perhaps constituted by it: the icon puts the habit in the *habitus*. If Humphrey Bogart's or Douglas Sirk's work now appears stilted or just socially weird, it's not because those films were free and authentic where today's Hollywood lets the genre make all the choices. Rather, those films look different because the gestural vocabulary of the age was in fact different, but not for a simple lack of strictures.

This is to say that any integration of parts into a whole, any coordination of behavior and appearance, any appropriate change in a part's function in response to a change in its quality, any *sense* whatsoever in the digital must be deliberately, intentionally put there using coded instructions to enforce determinate rules. The digital offers nothing for free; each of the trillions of amazing things that it can do is possible only because there is code designed to produce that precise output. And this principle applies also to digital input: every possible input to a computer must already have been anticipated in its general form. Clicking the mouse button has any effect in the computer only because there are algorithms that exist to tell the computer how to interpret as bits the electromagnetic signal generated over the wireless (or wired) connection between computer and mouse. A tool in the toolbar of a word processor does not somehow know what to do when it is clicked; it can respond to a click only because the software engineers who wrote the word-processing software included code to detect mouse clicks and respond appropriately.[2] If you are playing a computer game and you run past a wall with a crack in it, you may be able to hide in that crack, but only if some programmer has written code that determines how the crack can be used and how your avatar's body can occupy virtual space in that virtual world. Without such code, the crack might as well be painted on to the wall, for it can do nothing that has not been explicitly programmed, given a rule for its operation. (In fact, in digital spatial environments, textures are often flat two-dimensional patterns that are designed to appear to have depth, but really *are* effectively painted onto a surface.)

As the mutating cello and superficial crack begin to illustrate, digital individualization, which makes everything into a formal posit modeled on the bit, also restricts the sense of digital objects. In the actual, the sense of a thing, the reason that brings it about and makes it what it is, derives from its relations. Caught in a fabric of relation, an actual thing has many reasons, a whole universe of them, and making sense is a matter of choosing from those abundant reasons the apposite relations, the ones that suit the inquiry or provide the most illuminating explanation in a given context. In the digital, the severe isolation of each thing offers only a minimal reason, only those relations that have been expressly invented, exactly specified, and inscribed as code that reduces to rules of binary logic. If a digital object bears no inherent relation to its own parts, if those parts are not connected to each other in a complex tangle, then there is scant basis for sense; the sense of a digital thing can refer only to those rules that produced it. Why is that group of pixels darker than this other group? Why is this file saved in a new location? Why does this spreadsheet column show a much higher average value than the one next to it? Why are the corners of that rectangle rounded off? Inasmuch as we search for answers in the digital, we find always an algorithm, an encoded set of rules that generates precisely the output it is designed to generate.

But appeal to an algorithm is rarely a satisfying explanation.

People who work with digital machines may take for granted that the algorithm has behaved the way it was intended, such that questions arise not about the algorithm's operation but about the context in which it was executed. If a spreadsheet column has a surprisingly high average value, this is almost certainly because that column contains unexpectedly elevated values in general or because it includes one or two extremely high values that raised the average of the whole column, or possibly because someone entered the wrong formula for calculating averages. If a rectangle has rounded corners, it probably offers no explanatory value to note that the shape-drawing algorithm drew it that way; rather, a useful explanation likely refers to the fact that the user had the "rounded corner" setting *on* when creating the rectangle, or that the designer has used rounded corners throughout the whole project for a softer, more organic look. Such explanations appeal not to the binary logic that performs the calculations that produce the objects and events of the digital, but to the logic of representation that connects the binary to a world of human reasons and meanings. The logic of representation must respect both the binary, to which it reduces, and the actual, which determines its representational aims. As such, explanations rooted in the logic of representation remain close to the machine but traffic in representational possibilities; the sense of an object or procedure in the digital is circumscribed by the positive and definite ends that guide all digital processes.

The digital's rule-bound character highlights its ontological difference from the actual, for ruliness is the antonym of contingency. What happens by rule happens necessarily and exactly, which describes digital process universally. The implications of digital ruliness go beyond the illiberal discomfort of constrained behavior. That is, it is not just that the digital follows rules, but that those rules always reflect some deliberate intent: what happens in a computer is what has been designed to happen. Even the most complex and superhuman numeracy still proceeds, in the CPU, one tiny deterministic step at a time, following procedures that have been established to accomplish precisely those minute calculations.

Not only can the digital do nothing disallowed by its rules, but it also does *only* what the rules direct it to do by explicit instruction. It is this incapacity to invent or initiate that produces the poverty of the digital world; ruled by parsimony, the digital waits for more rules, doing nothing until it receives further instruction. Its world can be only as rich as the finite, discrete relations that were designed and implemented there. It thus follows a *principle of simplicity*: one gets out only what one has put in. This marks a key contrast with the actual, contingent world: contingency is the x-factor, the foment of the new, something more than or different from what was planned. Contingency is the font of creation, a guarantee of excess, whereas the digital gains its remarkable efficiency precisely by eliminating the scourge of the unpredictable and unruly.

In principle, one can always add another rule to refine or supplement software's behavior. Take the way we refer to time in casual speech. Human language relies on contextual clues to enable implicit evaluations of (otherwise) ambiguous time intervals appropriate to the situation. "In a while" likely indicates very different durations when it is a response to the question "When's lunch?" versus "When are you going to graduate?" A computer can be given rules to govern when it employs "in a while" or how it translates that phrase into a numerical duration using techniques of "natural language processing" (NLP). Crude rules will result in clumsy and sometimes inappropriate usage, but further rules can be added to take account of different contexts and exceptional situations. However, no number of rules can finally capture the infinite richness of the panoply of contexts in the human world that might be relevant to a given utterance of the phrase "in a while." The character of a personal relationship might tinge that phrase with sarcasm, such that it becomes a way of refusing to specify a time interval or a coy promise in return for patience. A previous remark in conversation or the shared knowledge of a book or film could provide the necessary reference for the contextualized meaning of "in a while," immediately and intuitively understood by the speakers. In effect, to guarantee that the digital machine can properly parse that phrase would require that the machine obtain a general linguistic competence. And while computational linguistics demonstrates that rules can simulate an impressive portion of linguistic competence, they are never wholly adequate, and NLP is most effective only when the context is assiduously delimited.

A comparison between a digital and an actual object dramatizes the digital's alternative ontology. Consider once again a fruit, noting now the contrast between actual and digital. An actual lemon, sitting on one's countertop, arises as a confluence of entangled forces from the microscopic (chemical reactions) to the cosmic (the material composition of the earth), from the mundane (a sale at the grocery store) to the grandiose (plant evolution as influenced by climate change). These forces enter into relations with each other, and these contingent encounters of forces—agonistic, sympathetic, allergic, punctual, resonant, parallel, interwoven, delimiting—work themselves out *as that lemon*. The lemon *is* this problematic encounter of many and diverse threads; it is the partial and passing solution to this billion-body problem. By contrast, a digital lemon would have a shape determined by one algorithm, a color by another, and a set of possible actions (and passions) given by still other algorithms (or data structures), and all of these algorithms and data structures are in principle independent of each other and of the wholeness of the lemon. It's easy to make a digital lemon purple without altering anything else about it. If its parts are coordinated, if the digital lemon demonstrates a coherence, it is only because that coherence has been deliberately assigned by some programmer, some rule-maker, who determined each part of the lemon, as well as their relationships, after the fact. A digital lemon is a collection of digital lemon parts and digital lemon qualities, and there is nothing about the digital lemon that guarantees any coordination among these parts and qualities.

It is contingency that places the actual beyond the reach of any simulation no matter how numerous or powerful. For any given actual event, contingency, "touching together," effectively ensures that *everything* bears some relevance: everything is in touch, however tenuous or distant. And this means that the digital cannot faithfully simulate even a single event, even the most formulaic happenstance of the actual, because it can never include in its calculation a tail so long that it finally circumscribes the universe. It considers only those anticipated and prescribed factors that are deliberately added to its code.[3] Again, the digital doesn't just fall short of the actual in some numerical sense, and it's not even a matter of the digital's necessary finitude versus the actual's concrete infinity, though this difference is already insurmountable. It is moreover that the digital is always planned, always predetermined, always intentional, even if those intentions are multiple and distributed and piecemeal. The finitude of the digital has a dramatically different character from the ubiquitous accident that determines (but never *finally* determines) the actual, because it lacks contingency. Where the actual has a plenitude of contingency, where it lives from contingency, the digital has the thin and barren determination of intention and necessity.

By its nature, the contingent cannot be simulated; any programmatic attempt to generate it, any ersatz, necessarily fails to be contingent, for contingency by its nature trips up every program, never falls into line. The digital can only multiply rules, whereas contingency is the potential to refuse any rule or to invoke a spontaneous and novel one. The digital can be numerically complicated, but what it will include in a given digital transaction is always already decided in advance, describable by formula.

Neural Networks

Neural networks, closely associated with the set of computational techniques labeled "deep learning," appear at first blush to offer a digital mechanism that overcomes many of the limitations laid out in this book's skeptical account of the digital. Instead of predetermined rules to govern their behavior, neural networks are trained, like Pavlov's dogs, through positive and negative reinforcement. Neural nets can be created using different techniques, but they are variations on a central theme: Initially the neural network is fed data and makes random (or rather pseudorandom) choices in response to those data to generate an output. If the output conforms to what the trainer considers correct or desirable behavior, then those (initially) random choices are reinforced so that the same data are likely to produce the same response the next time those data are presented to the network. Conversely, if the output is not correct (as judged by the trainer), those choices are deweighted so that a different response will likely be generated when the network is again fed those data. If the neural network is well designed and the training properly conducted, the network can eventually produce consistently correct responses, at least within the narrow domain on which it was trained. Neural networks can be trained to identify human faces (and other visual patterns), to produce appropriate responses to conversational or technical speech, to evaluate molecular chirality, to distinguish extraterrestrial radio signals from random noise, and to perform all sorts of other appropriate behavior in response to complex patterns.

Not only do neural networks thereby *learn* without being given strict rules in advance, but their internal complexity and use of pseudorandomness to generate output mean that it is often impossible, or at least highly impractical, to express the behavior of a trained neural network in rules even after it is trained and the weights of its statistical evaluations are fixed. Neural networks develop and employ an inhuman or extra-human knowledge, a micrologic of the actual world that reflects a vast complexity of subtle interrelations that become mean-ingful (or, rather, effective) only in statistical aggregate. The statistically reinforced pathways from the input to the network output may, through initially random reactions, select features of the input data that make no obvious sense to a researcher examining the self-constructed rules of the network; a network trained to recognize faces might base part of its recognition on a comparison of individual input pixels from different and seemingly unrelated parts of an image of a face, but this set of complex and unintuitive rules does not have to explain its methods in order to produce reliable results.

Traditional linear processing proceeds by assigned rules that reflect a theory, expressible in numbers and formulas, about the objects of analysis. A (traditional) algorithm for facial recogni-tion would need code that performs calculations on visual input data according to some reliable theory about the patterns that identify and distinguish visual data of faces and "recognizes" those faces based on a comparison of those patterns to pre-existing face data. Neural networks require no such theory be-cause, through training, they discover the patterns themselves, patterns often unrecognizable and even imperceptible to human observers. Thus, the assumption of a hidden logic is not only still a part of their operation, but moreover the core prin-ciple of their distinctive method. One applies a neural network to problems that exhibit complex behavior or patterns that are intuitively identifiable but difficult to articulate in rules, and the network develops it own rules as statistical relations among elements that cannot necessarily be understood in narrative or rationally transparent relations. But this works only if there is in fact an underlying logic, however complicated, one that can be represented by statistically weighted choices among different analytic criteria.

The founder of modern computing, John von Neumann, offers a direct statement of both the vast range of possibilities and the ultimate limitation of neural networks: "Anything that can be completely and unambiguously put into words is ipso facto realizable by a suitable finite neural network. Since the converse statement is obvious, we can therefore say that there is no difference between the possibility of describing a real or imagined mode of behavior completely and unambiguously in words, and the possibility of realizing it by a finite formal neural network" (310). To put it even more straightforwardly: anything that admits of positivist description can be adequately simulated in the digital. This axiom once again raises a key question of this book: what parts of our world cannot be described in posits?

In actual baseball, the batter attempts to hit the pitched ball. What determines whether this attempt succeeds? Of course, there is a statistic that tracks the frequency of a batter's hits, but this statistic is determined post hoc by the batter's performance and does not impose its reason on that performance.[4] We could describe the statistic in this case as *epiphenomenal,* meaning that it is a secondary effect, an artifact of something real (namely, the past record of hits and misses of that batter), but it has no causal force of its own. The statistical measure of a batter's hit percentage doesn't push anything around in the real world and certainly doesn't function as a rule or deterministic imposition on the batter's ability to hit any particular next pitch. For a given pitch, whether the batter hits it surely has something to do with how it is pitched. But "something to do with" is not a rule-based relation; if it were, pitching would be a bit easier, as one could use the rules to calculate the least hittable pitch.[5] The value of the batter's statistics becomes even murkier considering that the pitcher does not have perfect control of the pitch, for pitches, like the hitter's swings, are describable in statistics that do not influence but only reflect each pitch after the fact. Moreover, to refer the batter's success to the pitch is simply to move the problem of outcome determination back a step; one cannot say in advance of any pitch what might turn out to be relevant to its outcome.

Imagine the digital case: a baseball simulation that chooses, say on a probabilistic basis, from thousands or even millions of different *possible* pitches. Wouldn't the digital then be, in this simulation, just as complicated as the actual situation? And wouldn't it therefore effectively overcome the determinism of its digital rules by dint of an overwhelming number of possibilities? The premises of this hypothetical inquiry suggest, however, that, if the digital simulation is "just as complicated" as the actual case, it is nevertheless a different kind of complicated. In the digital, the relationship between each of those thousands of pitches and the batter's success is predetermined (at least on a probabilistic basis), a matter of rules that might be very complex and take account of many factors, but rules nonetheless. In fact those rules might amount to or even include explicitly a statistic that determines the odds of a certain pitch being hit by a certain batter, the statistic in this digital case governing the outcome rather than merely resulting from it.

One could possibly capture all those rules as a formula that summarizes as mathematics the code used to calculate the outcome of any given pitch to any given hitter. But, even if that formula includes many variables and a complexity that would defy unaided human calculation, still the formula itself attests to the rule-bound determinacy of the digital simulation, evincing not only the ruliness of the determinate outcome but also the boundedness of all the factors that could be relevant to that outcome. If a factor does not appear in the code, if it doesn't show up in the formula, then it can't possibly be relevant to the digital determination of the outcome of the batter's attempt to hit the ball. By contrast, the actual case places no bound on what might turn out to be relevant to the outcome of a pitch.

The different sort of complicated of the actual boils down to its inclusion of *contingency*. It's not that anything could happen; it's that there is no limit in principle on what could happen, what could be relevant. What the batter had for breakfast. How much humidity is in the air. Whether there is a clod of dirt stuck in the batter's cleat. The positions of the sun and moon. Whether a fan in row three yelled at the umpire on the previous pitch. It's not that these more and less proximal elements all matter; it is rather that you never know which ones are going to matter, that the actual world includes contingency as a well of potential in every event. There are many stories to be told about a given pitch and whether it is or isn't hit. But one cannot say in advance which of those stories will make good sense of the situation, for one can never anticipate the open-ended entirety of any situation. Except, that is, in the digital, where its entirety is hermetic, confined to the digital world, and thus not open-ended.

The baseball batter shows a truth of every event. If a particular reason (or a few reasons) seem like "the" explanation, that is only to the extent that one has already chosen a perspective from which to inquire. The 6 ball drops in the side pocket because it is struck a particular way by the cue ball; but while this is usually an appropriate account of the 6 ball's "behavior," it is by no means the only one. Some explanations might refer to the spherical shape of the balls, some might appeal to wagers on the game, some might mention the game's rules, some might cite the nap of the felt, and these are all reasons, determining factors in what happens to the 6 ball. Even the conception of the event as essentially *about* the 6 ball already privileges one set of perspectives among others, whereas another perspective might regard as relevant only who won the game, or whether it was completed, or whether someone cheated.

Zooming In

It is quite possible to simulate in the digital the zoom function of a camera, digital or film. But the usual caveats all apply: the simulated zoom necessarily operates according to a predetermined formula, which could potentially incorporate (simulated) imperfections or limitations in the optical elements of the camera, motion blurring, depth of field effects, bokeh, and so on. A simulated zoom as a checklist of formulas relevant to actual optical zoom yields a very good approximation of an actual zooming image. But because the digital simulation relies on pregiven formulas, even if those formulas have variables that can be changed from one instance of simulated zoom to the next, something always remains the same about the simulation. The digital zoom, like anything digital, is determined by a set of rules about how that simulation will be (mathematically) generated. One can introduce new rules that change the existing set of rules to introduce new effects or new relationships among the artifacts simulated, but then those *meta*-rules will once again establish a new bottom line that cannot be transcended. It is easy to see that yet another set of *meta*-meta-rules could make things still more complex and subtle, allowing the simulation to produce interesting differences in the basic algorithm for calculating the optical effect *and* interesting differences at the margins, margins of the image and also margins of nuance of the optical artifacts being simulated. But the rule that determines these variations of lower-level rules would then become a new rigid bottom line. The rules that generate the simulation will always be the same at some level; they will refer to some immovable frame of definition that determines what stays the same and what is allowed to vary each time the effect is simulated.

An actual zoom, using the same lenses, screws, and other physical parts, might well exhibit far less variation than the digital simulation with its hierarchy of rules to introduce visually interesting changes from one application of the effect to the next. But what nevertheless provides the actual zoom with a kind of auratic value beyond even the most complex digital simulation is the spontaneity of the rules. Actual zoom may display certain artifacts consistently, but every time that zoom is performed again, the precise combination of light, zoom speed, aperture settings, subject matter, lens imperfections, and every other connected element will produce a marginally different context, such that each zooming operation becomes a singular event, with marginally unpredictable results. That edge of unpredictability, that possible violation of any supposedly stable rule, is the upwelling of *contingency*, and in zoom as in most examples, it tends to manifest at the subtle edges, the subsensory or even unmeasurable margins of the actual, where positive effects do not submit to measure but are mixed together into what is often called *noise*.

Does the difference between digital and actual zoom matter? It feels as though for most purposes, the tiny, even undetectable difference doesn't really make any difference. An algorithmicist who spends enough time and effort to generate the most subtle effects from her algorithm may invest sufficient creativity to overcome (for practical purposes) its missing contingency, replace it with an ersatz singular creativity. But in practice, the principle of simplicity guarantees that even monumental efforts in the digital sooner or later reveal their limitations and demonstrate an aesthetic or ontological homogeneity. Digital simulation's bottom line is not just an abstract factoid about the digital, but is soon felt in digital production, which becomes stale or overly familiar and sheds whatever obscure creativity it once held in reserve. Thus there is constantly a need for a version upgrade, for a new algorithm, or a new rule to modify the existing algorithm. The upgrade culture that is a hallmark of the digital is not driven only by the economics of the high-tech industry, but also by the ontology of the digital itself, which cannot finally overcome an inherent monotony, a sameness that is the experience of bumping up against the digital's rule-bound bottom line. Frantic updates manifest an attempt to stay a step ahead of inevitable homogeneity, the dawning awareness of the limits of the current algorithm.

The ruliness of the digital is total. All manipulations of bits are determined by unambiguous rules, and those rules are themselves made of bits. Bits are the rules as well as the ruled. (This reflexivity reflects the digital's hermetic domain, illustrating that its world is closed unto itself.) If contingency is the potential to refuse rule, the de-absolutization of rules in general, then the digital's unfaltering respect for rules, its wholesale submission to rules, squeezes out contingency. The digital, for better and for worse, elevates rules beyond violability, and its regime is constituted and characterized by this absolute fealty, this unwavering commitment.

In Immanuel Kant's account of the rules governing the understanding, he acknowledges what Ludwig Wittgenstein later makes into a central concern of his philosophy: rules do not determine their own meanings and are never simply transparent, as though they maintained a direct connection with the logos. Rather, a rule must always be interpreted as to how it is to be applied. Kant assigns this interpretive responsibility to something like common sense ("mother-wit"), and in fact defines *stupidity* as the inability (or insufficient ability) to make proper interpretations of rules (268). Wittgenstein, concerned foremost with dismantling a naïve and overly rigid account of linguistic meaning, simply observes that, understood as a strict calculus of meaning, a rule always stands in need of another rule to say what it means. (He also acknowledges, along with Kant, that language and behavior generally don't work according to a strict calculus, and that we usually just understand, without reference to a chain of rules of interpretation; see §84f in *Philosophical Investigations.*) Nevertheless, the digital circumvents Wittgenstein's inquisitive critique, instituting rules that effectively police their own meanings. Because the digital is both doing and saying, it closes the gap between a rule's meaning and its execution. There is no interpretation, but only the action wrought by the bits coursing through the electrical circuitry of the digital machine. A rule is tantamount to a physical arrangement of that circuitry guaranteeing that certain input values will issue in certain output values. When bits express rules as commands or data, those rules too are just physically secured guarantees that certain outputs will follow from those bits being input into the chip. Digital rules are absolute. There is no room for interpretation, only a semantic inviolability underwritten by physical absolutes.

Sense-making or reason is thereby flattened in the digital. Sense in the actual is always the rescue of order from the precarity of disorder, a thread pulled from the tangled mesh that ties the universe together. (This hints at the ideological commitment behind Big Data analyses: they discover rules, trends, regularities, correlations, whereas the most potent meanings concern the anomalous, the outside, what happens beyond determinacy or predictability.) By drawing forth meanings from this overdetermined morass of reality, one makes sense of objects and events. The computer offers reasons, and one can trace a causal logic on any layer one examines; but the isomorphism of the layers ensures that all reasons collapse into the selfsame reason of determinate rules. There is ultimately only one digital reason, a single, formal sense to explain every object and every process, for it all reduces to the binary logic. The computer works the way it works because its bits drive it to, but they work the way they do because they serve a machine language and a coded logic of representation, and those codes exist because they allow the calculation of sound and image, statistics and simulation, work, play, and more, but always according to rules imposed as logical calculations to enable representational possibilities, determinate rules that establish general contexts, and await a realization when triggered by the appropriate inputs. The rules exist to accomplish possible calculations, unambiguous calculations of logical operations on 0s and 1s, which express a desire reduced to a set of determinate possibilities.

One might expect that, as ontologically disconnected, parts and qualities in the digital would tend to diverge erratically. But digital machines are not filled with purple spotted elephants or building-sized mosquitoes or software that executes random commands. On the contrary, the digital practically never deviates from its proper coherence, and parts tend to remain in perfect congruity with their qualities. But this congruity derives from the sparse and intentional relationality of digital objects and processes. Relations between a part and its quality must be deliberately and explicitly instituted in code, which can ensure their coordination, but which also implies that digital relations in general are thin, well defined, and highly finite. Because every digital relation has to be written as code, the number of relations is relatively low. The digital is a designed world, containing only what has been expressly and intentionally determined. There is no organic development, no accidental events or qualities, no surplus value that overflows the thing to which it is assigned. The digital conforms to the marketing slogan used to promote the graphical user interface (GUI) when it was introduced along with the original Macintosh in 1984: "What you see is what you get." It's a universe of stars, awesome in their multitude, but each so isolate that, collectively, they occupy no percentage of the vast night sky.

Granted, in the digital there are many layers of explanation of a given output or outcome. You could describe a result in terms of the flow of electricity through circuits, or in terms of atomic logical operations on bits, or in terms of larger flows and manipulations of data. Or you could look at the source code of the running program and describe the result in terms of subroutines and their inputs and outputs, or you could look at the interface and the software package and describe the outcome in terms of the way the software accepts input and produces output. *But the key insight here is that all of the digital layers are isomorphic, each tied deterministically to the others, such that there can be no substantive tension, no disagreement in these layers, and ultimately nothing available on one layer that isn't strictly captured in the others.* In summary, there is no point of entry for contingency, as every move in the digital is governed by an unambiguous rule. There can be no overdetermination in the digital, for every determination collapses into the single, simple determination of logical calculations on 0s and 1s.

Language and the Digital

John Cayley's chief objection to equating computer-generated word production with actual language is that language is an evolved ability, not an example of it-is-or-it-isn't positivism. Language is not just the production of a certain response to given stimuli, a response that could be codified or mapped in rules. Rather, language is a feathered domain of activity, intertwined with and in fact ontologically inseparable from all sorts of other things that people do. (In *Philosophical Investigations*, Wittgenstein famously emphasizes that linguistic communication relies on a shared "form of life.") We might say the same thing about many human activities, the entire range of human and even animal behaviors that puts labels on categories that don't admit clear definition: play, emotion, art, sensation. All of these things suffer a severe reduction when rendered as digitally simulated activities, so severe as to become wholly different sorts of things, pale imitations of their correlative human behaviors, produced by slicing off only a positive, empirical surface of an unaccountably diffuse and subtle aspect of human (or animal) being. But these positivist reductions benefit from the prevailing positivism of the current epistemologically privileged mode of thought. When representational semblance is the chief criterion, these reductions to digital code seem successful, because they are indeed based on representation and resemblance. That is, digital language simulation is generated according to criteria of how closely digital language can resemble the surface of human language, and our reigning model of reality identifies that surface with language itself, bracketing away the whole gamut of evolved behavior that is intimately tied to language. (This is somewhat akin to Wittgenstein's discussion of private language, in that private language seems possible only when one forgets all the things that language is and treats it instead as a positivist phenomenon. If you disregard that language is also gesture, context, sympathy, community, and so on, then language becomes merely the possibility of generating statistically likely sequences of words, which is precisely the way that word-vector databases, like many other digital NLP techniques, are organized.)

To emphasize that language is an evolved phenomenon is to note that language is intertwined and continuous with a range of capacities and behaviors, not an isolable skill. If we have the capacity for language, this capacity gets exercised in many different ways, not all of which are commonly acknowledged as linguistic behaviors. Does language already imply some sort of mathematical or logical ability? Does it imply an aptitude for abstraction? Does it already include some sort of musicality, a certain possibility of a relationship to temporality, a mode of community, a relationship to space? These are unwieldy and ill-defined questions, but they can be answered affirmatively with some confidence in each case, for only a moment's thought reveals that language is deeply embedded in a wide range of associated and disparate behaviors.

This range of paralinguistic capacities, imagined perhaps as a cloud of thoughts and gestures rather than a background against which language happens, presents another aspect of contingency. In this case, the interdependencies are so dense—the range of related phenomena diffusing into nature, history, being—that language can be said to be born of contingency, to depend essentially on the coordination of uncountably many and diverse factors.

This is not just another *zombie* question. In philosophical discourse, a zombie is a being that acts just like a person, is in practice indistinguishable from a person, but by definitional hypothesis has no inner life, no consciousness. Some philosophers believe that all of us are, according to this definition, zombies; they believe that our sense of having an inner life or consciousness is illusory, a story that our zombie selves narrate to ourselves. Others propose that we have no way of knowing whether anyone, besides oneself, is a zombie, and so *everyone* else might in fact be a mindless automaton. In any case, this discussion of language might seem to resemble an argument about zombies: if computers can effectively simulate language behavior, if they can accept linguistic input and generate appropriate or human-like linguistic output, then the question of whether they "really" have language or not is moot. Indeed, the most well-known criterion for deciding whether a digital device should be considered intelligent, the Turing test, is tantamount to this question of zombie language.[6]

The notion that language is evolved also seems to exclude digital language in another related way: inasmuch as it is evolved, language cannot be understood as the sum of independent parts. Evolution is a process of trade-offs, balances, progress and regress, tying together a huge number of different but connected phenomena to make a meshwork of a creature that cannot sensibly be separated into parts. Any such separation is a post hoc cleavage that also severs from the disconnected parts the whole (and its sense) that is greater than their sum. Evolution operates synergistically, tying together elements that were never really distinct to begin with, and tying them so closely that they can be distinguished only artificially. One sees this at work in the development of a fetus, in the structure of the nervous system, in the proprioceptive essence of bodily movement. The systems of an animal are the results of complex evolution that does not allow a piecemeal division that would still honor the meaning of the whole.

Even evolved qualities seemingly simple and separable hide an unsimulatable complexity. Consider eye color. To have gray eyes might seem like a simple enough thing to simulate, and indeed one can easily instruct a computer to apply a gray color to the irises on an image of a person's face: *voilà*, gray eyes. But gray eyes too have a complicated evolutionary history, one that connects them genetically in variable relationships to other aspects of a person, including skin tone, hair color, and still other traits that have not been adequately studied, including possibly behavioral factors, various cognitive aptitudes, phenotypic tendencies, and so on. Though there are two genes that exert a primary influence over eye color, it is not known how many genes may have a secondary influence, and each of those genes may also affect other parts of the person. The existence of certain alleles in a person's genetic code already bears a complicated relationship to the expression of associated characteristics, as the genes must be not only present, but activated, and the proteins that are produced on the basis of that genetic material have diffuse effects throughout the organism, including phenotypical effects such as gray eyes.

But as has been demonstrated above, the digital has only parts to make a whole, and only specifiable, definite relations among those parts to produce a sense of the whole. Digital synergy would be by design, not by evolution, and it seems impossible to capture the multiscalar complexity of synergized animal life by adding parts together.[7] Any encounter or meeting or grouping or addition of two digital parts has to have been anticipated in advance, its results outlined in those parts or in some further code that oversees combinations of parts. (And if there is an overseer, this still regulates the parts themselves to some extent, for they have to have hooks built in so that they can be manipulated by the oversight algorithm, which can operate only on properly conforming parts.) It would be more or less accurate to say that the digital has only parts, as even the relations among these parts effectively act as parts. (Note again that the piecemealization of the digital applies not just to data but also to action: a given task, a given possible behavior or procedure in the digital must be broken down into parts, themselves broken into smaller parts, until one finally reaches a description of the action as sequences of discrete logical operations on 0s and 1s.)

Many an apologist for the digital would argue that its missing contingency doesn't matter, given that its numerical plenitude—the number of options it can consider, the speed of its calculations, the vastness of its storage, the variety of different inputs and simultaneous processes it makes available—amounts experientially to much the same thing as contingency. This is *to substitute an epistemological limitation for an ontological openness.* A person confronted with twenty different options feels much the same, it is said, as a person confronted with a million or even an infinity of options; in each case, there are too many to evaluate, so the ontology of contingency loses its distinct potency in the face of the epistemology of multiple determination. Plenty of those who interact with the digital claim to experience no particular restriction; they feel that the digital at every turn offers more than one bargained for, that it generates a surprising excess. Because it can take in so much data, because it confronts the incomputable, because it offers more than one can even imagine, the digital yields results that are not only fundamentally unanticipatable (short of running the same program with the same data), but genuinely creative. The digital carries out operations that no person ever could, and so gives us something extrahuman.

Extrahuman indeed, but ontology is not finally trumped by epistemology, for one feels the difference between digital and actual, the specific difference made by contingency, over and above the limitations of hardware and resolution. This feeling may go unrecognized and unacknowledged, because we are encouraged to regard our experience, on and off the computer, in digital terms. But to know that an outcome has been planned, to know that it is the result of intentional processing, even if those algorithmic processes were designed by many subjects and even if one cannot oneself predict the outcome, to know that the outcome has been destined under the pulsion of an always instrumental intent, is to experience the finitude of the machine, the sometimes comforting assurance that the result has already been thought, possibilized, laid out as a target of desire. One knows in the digital that no situation can be unthinkable, for every situation has been inscribed within the limits of thought. The machine always behaves itself, which is precisely what makes it so attractive, but also ultimately so boring. It can never give back more than or other than has been invested in it. Its creativity cannot exceed that of its designers; it will do whatever is instructed, a machine of your desires, but it will do no more than is instructed, a machine of rigid necessity. By contrast, the creativity of the universe is all-consuming, entire. Each moment of actuality is the birth of the wholly new. And thought can never reach this apex of creativity, but can only give itself over, prepared to be taken up in this creative foment, and so moved to thought's highest achievement.

The Self as Posit

Selves attract much positivist assessment. For a century and a half, governments (and other institutions) have leveraged statistical representations of populations to control the distribution of goods and services, to track trends in the movements, desires, and habits of human beings, and in brief to make strategic choices about who lives and who dies within a society. The use of abstracted data to manage life in populations is the core practice of *biopolitics*, an understanding of the relationship between social institutions and individuals made famous in the critiques of Michel Foucault. These statistical methods treat individual people as comparable or even equivalent to each other, setting aside individual difference and making use of broad categories to lump together groups of individuals by age, gender, race, economic status, or another demographic determination. This is a classic positivist outlook, in that persons do count, but only as *yet another*, an increment of one to the tally of population or category. Personhood is reduced in these statistical analyses to the mere fact of existing, human being captured in the simplicity of a posit.

The positivism of biopolitics continues today unabated, but it has been to an extent reformatted by an alternate positivism born of the digital age. Particularly in online settings, individuals become data sources from which are drawn numerous bits of information, each of which may or may not be significant on its own. Those bits can be collated to constitute a data profile of the individual, but even partial data, when grouped together, make possible a statistical regression that determines statistically significant groups and calculates a degree of membership in each group for each individual. These statistically constructed groups substitute for the broad categories of identity employed in traditional biopolitical manipulation; there may be no legible significance in a group of individuals who all watch television in the early evening and also tend to buy the same brand of laundry detergent, but such statistically determined categories allow marketers and political campaigns to identify likely marks, offering to those individuals not only a product or perspective but also a sense of belonging to some group. The common practice of pitching to the user additional products or media she will "probably like," based on an analysis of what people statistically similar to her have liked, is tantamount to offering her participation in an identity category. (Research on the Netflix selection algorithm reveals that Netflix has over two thousand "taste communities" into which it sorts its subscribers [Pajkovic 2021, 4].)

Though the reduction of persons to aggregate but partial statistics and the collection into categories by way of these statistics are both artifacts of biopolitics, the new digital order introduces some important differences. One difference is that the digital data profile no longer expresses the schema of an overarching institution such as a national government, but instead represents a collection of perspectives, none of which claims to capture the entirety of the person, but only certain utilizable facets. A second point of departure from traditional biopolitical power dynamics is that the individual is not simply accounted from above, but willingly (though not always wittingly) provides the information and even welcomes its collection and re-presentation, often regarding it as a service, like personalization or efficiency improvement. Many people gladly assent to their webpage visits being measured and clicks being tracked in order to ensure pop-up advertising more relevant to their specific interests, for example.

This digital profiling and analysis is still a positivism, inasmuch as data are always positive: a numerical (or categorical) representation of a particular trait or fact, standing alone as an assertion about the world. The individual person has been broken down, and is available to be re-atomized according to new categories, so that the posit no longer circumscribes the individual, but rather sets forth her various aspects subject to measurement and comparison. This positivism therefore exists in tension with its own fragmentary, diffuse, and pre-individual loci, as though the posit of an individual person, still the object of biopolitical capture, has been taken apart and reconceived as a privileged nexus in which various data accumulate. Positivism's scale has adjusted to the scale of the digital, which establishes wholes only as collections of parts, themselves collections of parts, such that a person online amounts to a long list of data points that manifest complex interrelations but also underlying independence, as though a person were an accumulation of habits, desires, sizes, tastes, lifestyles, credit-card debt, marital and family status, and other categories. From the perspective of these data buckets, there is no whole person anymore, as the recruiters, advertisers, and political campaigners are interested only in the personal data relevant to their operations; Gilles Deleuze (1992) tracks this breakdown of the person as a shift from the individual to the "dividual," from an integral whole to a collection of manipulable and calculable parts.

Could this list of aspects ever be adequate to a person? Would a positivist account of your self, your person, ever feel as though it had done justice? Can you imagine a description so complete that it captured who you are? Is there a description of any length, at any level of detail, that would accurately reflect your self, your being? Surely we have reason to doubt the possibility of a truly adequate account of the self made of bits and pieces; one's being, a person's way of being, is an outstanding example of becoming, resistant to positivist capture. At the level of formal ontology, one might point out that one's self is definitionally altered in the very act of evaluating the possibility of capture. That is, it is not the same self that examines the list of positivist claims about itself as the self that decides on the adequacy of this list. And any given statement is surely inadequate to its lived expression in a self, which always adds an uncapturable sense of self (the singularity of being *there* that Martin Heidegger calls *Dasein*) as the background against which any particular trait is expressed or known.

Though contingency is immune to simulation, it is possible in its absence to mimic some of its effects, at least temporarily and within a circumscribed domain. Digital media reveal their ruliness or lack of contingency often only slowly. Not only does a given piece of software offer overwhelmingly numerous possibilities, but software undergoes frequent updates, some incremental and some more dramatic. And it is sometimes possible to switch to a new piece of software in a related category to open up still more possibilities. Even with one piece of software, the possibilities available are often too many to experience in a lifetime of computer use. Does the staggering number of possibilities, or the incremental novelty of an update, or the more dramatic switch of an upgrade, or the still more significant shift of new software substitute for contingency, mitigating its lack to the point that, at least experientially if not ontologically, it doesn't matter?

Luciana Parisi proposes in her 2013 *Contagious Architecture* that the sheer numeracy of the digital does indeed dwarf its straitened ontology, such that its world is even more unpredictable, creative, and unruly than our familiar actual world. She depicts an ontological structure of the digital resembling the messy and tangled mesh set forth in chapter 3 of the present book as an account of actual ontology. Supporting her vision of the complexity and supple richness of the digital is the assertion that, in their networked contact with each other and in encounters with unanticipated inputs, algorithms are exposed to "infinite" sets of data, an infinity and variety that could not possibly have been anticipated by the designers and programmers of the algorithm. This is an image of the digital as a florid jungle of vital algorithms, no longer caged within their intended domains of operation, but set loose to encounter unexpected others and wild input data. New algorithms are born of these encounters, engendered by the perverse intercourse of algorithms and data, with human engineers serving as digital midwives.

The fecundity of the algorithm is possible, says Parisi, because algorithms all hold within themselves a genital nugget of unknowability, essentially a potential for contingency. This contingency is available in any given algorithm, awoken when exposed to unanticipated contexts, but its productivity is really unleashed when we consider the way that the whole digital domain develops as new algorithms are generated in relation to preexisting algorithms and new possibilities. Parisi situates this untamable secret of the algorithm in an ingenious artifact of the mathematics of computation, a number invented (discovered?) by mathematician Gregory Chaitin that, with no great modesty, he calls "omega." Briefly put, omega is the probability, expressed as a number between 0 and 1, that any given algorithm will eventually halt; it represents the chance that an algorithm chosen at random from the set of all possible algorithms will eventually terminate. As mentioned above, Alan Turing, building on Kurt Gödel's work in logic, proved early in the digital era that there can be no general procedure to decide whether a given algorithm eventually comes to a halt or does not. Consequently, it is assuredly the case that omega is, in a sense, incalculable. It is possible to calculate the first n digits of omega for any finite n, but, as Chaitin emphasizes, almost nothing can be known (or reliably surmised) about the value of omega past the brute-force calculation of its initial digits. In fact, omega represents a mathematical apex of unknowability: it can be proved that its digits can never manifest any pattern, in which sense it is maximally random.

Omega's patternless unknowability bears striking similarities to contingency. Until, that is, one recalls that contingency and randomness are not at all the same thing. Like contingency, omega rejects the authority of any transcendent rule (except the constructive rule that generates its successive digits); it refuses any pattern, and the only rule to be discovered in the examination of its successive digits is that they abide by no rule. But, while contingency is defined as the potential to transcend any particular rule, it is not tantamount to the arbitrary or senseless intervention that maximizes unruliness. On the contrary, contingency puts everything in touch with everything else; starting at any point, it traces its meshy relations to the ends of the universe, ensuring an overdetermination, a multiplicity of sense. Chaitin acknowledges omega's interesting but ultimately false claim to contingency: "In pure mathematics all truths are necessary truths. And there are other truths that are called contingent or accidental like historical facts. . . . And whether each bit of the numerical value of the halting probability Ω [omega, represented in Chaitin's discussion as a binary number] is a 0 or a 1 is a necessary truth, but *looks like* it's contingent" (138; emphasis added). Bound by the same sort of inexorability that propels algorithms to their inevitable conclusions—or nonconclusions for algorithms that do not halt—omega's sense is the formal necessity of a statement of pure mathematics. Its pseudocontingency does not reveal a deep sense through the connectedness of the universe, but only a caesura in the image of mathematics given to thought.

Parisi's appropriation of Chaitin's omega thus does not succeed in investing the digital domain with an independent creativity. Every data structure in an algorithm has already been anticipated in outline; every possible input must be inscribed as a shadow, a form awaiting its content, even before the algorithm is executed. And those possible inputs are ultimately finite, even if their staggering cardinality tempts us to imagine there an infinity. Parisi rightly recognizes that the sheer numeracy of the digital produces an overwhelming complexity (of a sort), but wrongly surmises that this epistemological excess overcomes the ontological sterility of the digital. The digital domain does indeed change and grow, but it does so only through the ongoing creative input of engineers and programmers. The creative element must always be donated from outside the digital, for the digital can offer only as much as it has been given.

This rulebound, anticontingent behavior of the digital machine, in which one gets only what one has instructed the machine to do, might sound restrictive and laborious. But those hard limits on what the machine can do also grant it its efficiency, breadth, flexibility, reliability, and all of the fantastic powers of the digital that have made it, in short order, the overwhelmingly dominant technological paradigm. As an essentially individual posit, every part of a digital object can be independently manipulated, copied, moved, or otherwise altered, providing an extraordinary flexibility and exacting control. Because there is no inherent relationship between one pixel and the one next to it, an image can be adjusted to exact specifications with perfect precision, leaving everything else entirely undisturbed. A digital visual artist can alter a single pixel at a time, nudging its color almost imperceptibly darker or more blue, without any effect on the pixels nearby. A traditional painter with a material paintbrush has no such precise control, and must accept the accidents, the extra unintended effects that haunt even the greatest technical ability, sometimes serving as the very source of creative possibility. The super-fine digital control, the exposure of every part of the object, the possibility of massaging every tenth of a second of sound in a recording, is incredibly powerful and markedly different from pre-digital modes of production.

A calculated and subtle relationship can be written in code to ensure that an enlarged digital cello makes a deeper digital sound or that a rotten digital lemon makes a juicier splat than does an unripe lemon when smashed with a virtual mallet. (An actual cello, bearing a myriad of internal and external relations, typically changes its entirety, at least subtly, in response to any small alteration; replace one string on the cello and the tension in the other strings also changes, as well as the slight curve of the neck, the tightness of the tuning pegs, etc.) Such code to coordinate behavior and appearance, to make sure the parts of an object are interconnected appropriately, is an essential mechanism of adequate simulation, and in fact the execution of such rules in code is basically what computers do. Code, constructed according to a logic of representation, establishes rule-bound relationships among various objects, their parts, their appearances, their behaviors, so that they act in ways that are useful or entertaining or informative or predictive or beautiful. And those coded rules work because there is no contingency to disrupt their operation, because, even though the machine may execute billions of logical operations in a single second, every one of them produces exactly the outcome it is supposed to, a self-contained world of satisfied desires.

Designed around perfection, based on an element that behaves like an idealization of itself, equal to its own concept, the digital machine achieves a nearly flawless efficacy, but by the same token works much less well when one does not know exactly what one wants. Action in the actual frequently proceeds by a method of trial and error, where one takes a first step and then recalibrates before the next step, and recalibration may alter not only the heading but also the goal. In the actual, we invent as we proceed, relying on the ubiquity of contingency to ensure that every step confronts something novel. This is easiest to observe in practices of art-making, but it is just as applicable across the spectrum of human activity, from cooking, to child-rearing, to athletics, to language learning, to meditation, to throwing a party. Improvisation is the rule of action, as each moment brings about something new, at least at the edges of the frame, and one must constantly renegotiate one's relationship with the world. The digital's perfection is the minimization of this essential element of existence. To take a step in the digital is to hold a steady goal and chart a direct path toward it. Any gesture is strictly correct or incorrect, as measured by its approach to the goal. Intention in the digital can fill out the entirety of an action, in that there are no accidental effects, nothing that is not part of the programmed consequence of an anticipated input.

This is not to say that experimentation in the digital is impossible. One can enter a parameter value without knowing what effect it will have or move the mouse idly and without deliberate intention or choose from a list of sound samples without knowing how the chosen sample will sound within the overall composition. Such techniques of self-imposed not-knowing may begin to simulate the experience of accident in the actual, but it remains the case that the outcome of setting a parameter or moving the mouse casually has the exact and predictable effect that the system of hardware and software necessarily determines. A person can employ a digital tool in a limited practice of trial and error, but the digital contributes no error, and so leaves all of the contingency on the side of the user.

Perhaps this account of digital perfection is too absolute. There may be some room, however slight, for a degree of contingency to enter the digital machine. When the machine errs contingently— when a digital machine fails to operate according to the rules that normally govern its digital procedures—it is usually a case of hardware that either writes or reads one or more bits incorrectly. Sometimes a piece of dust alights on the surface of the chip, or the magnetic substrate loses its magnetic sensitivity after too many read/write events, which might interfere with reading or writing bits.[8] Typically, a bit that is read incorrectly yields a numeric voltage, reflectivity, or magnetic-field strength outside the normal range of acceptable nominal values for either 0 or 1. That is, a dust particle doesn't usually trick the machine into thinking that a 1 is a 0, but rather results in a failure to read the bit properly. In such cases, the logic system controlling the reading (or writing) of bits tries again, sometimes multiple times. If multiple read attempts fail, the system can often reconstruct the missing data by calling on integrated backup systems: some storage media include full or partial data redundancy in the basic storage logic to prevent errors. Other systems store general information about the stored data, and it is sometimes possible to reconstruct missing data using this general information. *Checksums,* for example, add up the values of various numeric data and store the result along with those data. If the data are read back but do not add up to the stored checksum, then the system recognizes that data have been incorrectly retrieved and in some cases can fill in the missing data by calculating the shortfall relative to the checksum.

Failure to read a bit properly is a relatively rare occurrence, usually detectable and often correctable. When it isn't correctable, the system may generate some sort of error message and a human being must intervene to fix the problem, say by finding a backup of the corrupted data. Even more rare are those occasions where the system reads a 0 as a 1 or vice versa, such that the data (or commands) are processed but not as intended. Changing a single bit in the representation of a machine-language command usually yields a different command altogether, while changing a single bit of a single datum can alter a numeric value significantly or make an alphabetic letter into a different one or change the color of a pixel, and so on. But digital systems, with their hard edges and absolute assertions, are quite brittle, at least in cases of mistaken data or commands, and the most typical consequence of a 0 mistaken for a 1 is that the machine crashes. This can happen for a wide variety of reasons. A misread bit can create a command code that refers to no command, such that the machine balks, or it might cause the machine to lose track of where in memory to find the next command to execute, such that the CPU tries to execute a stream of commands read from a memory location that doesn't even contain program code, analogous to a person in step 4 of the instructions for building an IKEA cabinet suddenly starting to read her next instruction from the middle of a Dostoevsky novel.

The point is that digital systems are designed around their brittleness: digital errors are generally caught before they cause trouble and are often fixable even without external intervention. When errors do pass through the digital system, they rarely result in a serendipitous surprise, but instead just cause everything to break down. Such errors do evince a contingent occurrence in the digital, but unlike the accidents and marginal resistance one encounters in the actual, this digital contingency is rarely an opportunity for creative reorientation or innovation.

Maybe human error in the digital domain can be more interesting. If a doctor enters the wrong value into the "Age" field of a database record of a patient, or if an artist's hand slips as she moves the mouse to draw a line in an illustration program, or if a programmer forgets to include a line of code to increment the value of the variable that keeps track of how many times a given subroutine has been executed, the results seem at least potentially generative if not necessarily desirable. The doctor's mistake might contaminate the results of research conducted using that database, leading to unfounded conclusions and an undeserved award of grant funding. The artist might discover that she likes the jagged look of the image with the line in an unintended place, and it could even constitute a break from her usual style, one that she seeks to replicate in future projects. The programmer could react with bemusement when the program runs, as the subroutine with the missing line never reaches its terminal condition and keeps allocating additional memory until the system runs out of memory and the machine crashes.

Such errors are not strictly preventable and are undetectable by the digital system; that is, the digital system has no reliable way of judging whether a given action by a user is intentional or accidental. Instead, digital systems are designed to follow rules without question, especially commands entered by their users. Executing commands unfailingly is the great strength of computers, even though those commands might sometimes be accidental or misguided. Digital interface design occasionally attempts to mitigate the potentially disastrous slavishness of the digital machine by requiring confirmation before undertaking consequential actions, effectively demanding to receive the same instruction twice: "Are you sure you want to empty the trash?" or "Reformat disk?" More rarely, a digital software system might keep track of simple patterns in user input, alerting the user when the pattern is broken: "You usually pay this merchant less than $100. Are you sure you want to enter a payment of $1,000?"

Human error is more common than machine error and more likely to generate happy accidents. Still, the digital's great advantage of nearly perfect instrumentality works against it as a context for creative happenstance. For, even the contingency of human accident is minimized by design in the digital: digital systems are designed so that human input is mostly deliberate and intentional, emphasizing conscious and knowing actions and opening few channels for more intuitive or unconscious behavior. Mouses and keyboards are relatively narrow mechanisms for user input, accepting only certain kinds of movements and discrete presses of keys, such that a user can maintain a complete control over the instructions she issues. Compare these forms of control to riding a bicycle or playing a viola, where there are many things to keep track of, as one's whole body is involved and one must entrust much of the control to habituated actions; a cyclist does not have to think about shifting her weight during a turn, and an accomplished violist does not have to decide in advance to lift her finger off of the finger board, for these actions are automatic even as they demand significant subtlety. Typing on a keyboard and using a mouse no doubt also become largely automatic and thoughtless behaviors with practice, but they remain narrow bandwidth, refined mechanisms, allowing a complete control and minimizing accidental or unintended inputs.

Foregoing contingency by design, the digital machine satisfies our desires, as long as those desires conform to the positivism of the digital. But it turns out that a great deal of the world can be captured by positivism, almost everything, from some perspectives. What escapes positivism is only what is not really a thing, what does not fully assert a being; one might even say that what positivism cannot capture is only *what is not,* not an entity, not a thing, not present, not there. Much of the actual, if not its most essential and constitutive dimension, can be captured in posits. A representation of reality as individuated entities with secondary relations proves not merely adequate, but mostly just as good as the real, particularly when measured by its instrumental value. As has been emphasized repeatedly in the preceding pages, posits subject to manipulable rules make for a remarkably efficient and effective means of addressing the world; as long as one's desire is itself positive, the desire to realize a distinct possibility, that desire enters the reach of instrumentalized reason. What makes digital systems so attractive is that they induce in their users the same kinds of problems that they can solve.

This world of guaranteed solutions sounds both thinly superficial and temptingly rewarding, and both qualities derive from the same constitutive factor, a lack of contingency. Because digital ideology, prevalent even before the growth of digital technologies, prepares the way for those technologies, the digital's benefits seem effortlessly to trump its costs; who needs contingency when one can get whatever one wants? The digital offers a world stripped, at last, of the irreducible, the remaindered, the unassimilable, the resistant. In a world rendered positive, contingency doesn't even show up, and is thus sacrificed without a thought.

But the appeal of the digital goes beyond just this cost–benefit analysis, for even its apparent cost also presents itself as a benefit. A world of necessity, which might threaten one's sense of freedom or autonomy, seems instead to offer a desperate comfort, an assurance that everything is in order and indeed that there is an order. As the twentieth century heralded the erosion of master narratives—the atrophy of religious faith, the submission of the national to the global, the destabilizing of family and community—leaving much of humanity uncertain of its place or reason, digital technology erected a world that returned to us an absolute reason, an accessible solution to (almost) every problem, a universal language, and the existential assurance that everything finally fits together, that it all makes sense, and that one can always "Undo" any misstep.

This analysis has finally led back to the question in the opening pages of this book: What does the digital do? And we might reframe the question as: What desire calls forth the digital as technology? The desire that motivates the digital is not just rationalism or positivism, which have plenty of nondigital versions, nor is it the fetish of numbers. Rather, the underlying desire is *to live in a space of necessity,* where everything that happens happens necessarily. For the world to make sense, for the world to submit itself to sense-making, is for every event, every object to exist according to a rule, for rational sense-making is most assured when it is a matter of rule-following. The elimination of contingency, even when revealed in the bright light of analysis, is therefore taken not as a deficit, but as a great strength, the return of order, the satisfaction of sense.

Enticed by the digital's reassuring order, users embrace digital values as they employ its machines in more and more parts of their lives. Because users get what they want, because the digital keeps its promises and does what it is supposed to, those values come to offer evidence of their own correctness. Digital efficacy validates the ideology that it represents, effectively spreading that ideology beyond the immediate context of digital devices and into the worldviews of its many users. When using the digital, one adapts one's desire and one's process to the positivist lack of contingency, choosing one's goals from the multitude of available digital possibilities. Having acceded to this regime of choice as a substitute for creative freedom, and having been rewarded with a rapid response to one's desire, users adjust their extradigital desires in like manner, coming to see a life even in the actual as a set of individual choices as though selected from menus.

This illustrates the coercive charm of the digital: prepared to solve any problem within its positivist domain, it urges its users to conceive positive problems, problems whose solution is the realization of a possibility, problems that are not about creative reformulation or experimentation or appeal to a new frame, but rather about means and ends that are already available. The digital can help you go where you want, as long as where you want is a specific place, an identifiable endpoint. It is an immensely satisfying and comforting conservatism: as long as you stick to what is available, foregoing any radicalism that reaches beyond the determinate and the possible, you may have exactly what you wish for. No wonder the digital provokes such intense desire and such zealous commitment.

Thus does the digital recast the world in its image, replacing gods and kings with its verifiable ubiquity, its liberal neutrality, its unerring correctness, its claim to futurity, and its role as the source of our desires and the authentic promise of their satisfaction. The digital can represent only what is positive in the world, and so teaches those bathing in its aura to recognize the world as made of posits, amplifying and rarifying the positivism already latent in our world. We don't ordinarily encounter color as a set of precise distinctions among shades or brightnesses, nor visuality as a congregation of points in a plane, nor people as sets of categorized characteristics, nor creativity as a choice among available tools in a toolkit, nor thought as a set of discrete steps from obscurity to enlightenment. But the digital selects those parts of our lives and then, inserting itself as a model of the actual, enforces that selection as a valid way of encountering not just the digital but also the world it represents. Because the digital is always a model (representation), it makes an implicit claim to validity, presenting itself as comparable to or similar to the actual world in important respects. The next chapter considers more closely how the digital trains us, its users, to see the world as made in its image.

◀ **6** ▶

What Does the Digital Do to Us?

In an analysis of the notoriously addictive computer game *Civilization II,* Ted Friedman suggests that gameplay can be transformative, altering the player's relationship to herself and her world: "The pleasure of computer games is in entering into a computer-like mental state: responding as automatically as the computer, processing information as effortlessly, replacing sentient cognition with the blank hum of computation. When a game of *Civilization II* really gets rolling, the decisions are effortless, instantaneous, chosen without self-conscious thought. The result is an almost meditative state, in which you aren't just interacting with the computer, but melding with it" (136–37). Friedman describes the process by means of which one achieves this "computer-like mental state" as a "cybernetic circuit," wherein player actions prompt responsive feedback from the machine, which in turn triggers further player actions. Importantly, this circuit is not only cognitive; instead, what circulates from player to machine and back is simultaneously affective. Friedman's description implies that, well beyond a mood, the player develops a new *way of being.*

Many digital games are designed to encourage this transformation of the player into a semiautomaton, training her to construct a vocabulary of automated input gestures to deploy in response to the recurrent actions of the machine. This may happen at different scales, depending on the nature of the particular game. Twitch games require rapid and properly timed responses; a forward lunge by the machine-controlled "enemy" may be countered with an immediate backward leap by the player, giving the player a mere quarter of a second to press the correct buttons and avoid damage. Success in such a game is a matter of training oneself to the point that these rapid responses become automatic and unthinking, as though, to follow Friedman, the player were herself a computing device that acts without thought. A game like *Civilization* (and its many sequels) gives the player as much time to make her moves as she wishes; control does not return to the computer until the player presses the "End Turn" button. But the basic structure of training is similar to the case of the twitch game: the player observes the tactical choices of the (rule-bound) machine-controlled opponents and executes countertactics developed because they have been more or less successful (or otherwise rewarding) during prior play.

Friedman's essay opens with two examples in which the player's transformation extends beyond the game and into the actual world. One is a television advertisement depicting a kid on vacation who has been playing so much *Tetris* that everywhere he looks he sees tetrominoes, blocks of tiles ready to rotate and slide into place. The other example is Friedman's own experience in the arcade of playing the motorcycle racing game *Pole Position* and then finding himself hugging corners and "darting past cars" while driving home in his car. These experiences, familiar to avid gamers, offer a stark demonstration of a phenomenon that, while especially vivid in gaming due to its affective intensity, is common to all digital interaction generally. To experience the digital is to internalize its ways of being, its ontology, and then, often and unconsciously, to carry that ontology out into the wider world of the actual.

This is not a description of an insidious force built into the digital, brainwashing those who come into contact with it. On the contrary, the dissemination of digital values, though no doubt bolstered by the digital's compelling efficacy, follows a pattern common to much human activity, especially the value-laden activity of learning. To learn is to develop a habit, a way of moving, of thinking, rhythms and flows that one repeats, adjusting one's habit to accommodate changes in the situation with which one is confronted. To learn to play the trumpet or use an arc welder, one finds a way to fit one's body, one's intention, to a process, sometimes guided by a desired outcome. When the learned activity fundamentally engages the material world, the ubiquity of contingency guarantees that each encounter is subtly different, and the acquired habit is not merely a rote reproduction of past experience, but a renegotiation that demands of the learner something new every time. Learning in the actual is therefore an endless process, for one must learn anew each time one undertakes that "same" activity.

To play an instrument or to use an arc welder is always to make vital adjustments in real time, to adapt to the circumstances of the moment that are subtly different from previous circumstances: one's fingers are stronger, there is a nearer deadline, someone else has used this instrument or tool in the interim, you feel distracted after a heated argument with your good friend last night. The learning, the habit, serves as a foundation that provides a sufficient stability so that one can then experiment or invent or problem-solve in the marginal domain that habit does not fully inhabit. And it isn't clear where this margin would be found in the digital, for the digital has no margins, another meaning of its discreteness. It is what in topology is called a "closed manifold," one that contains all of its own limit points. Digital objects and events have no margins because every thing, every point is either definitely in the digital, part of the object, already accounted in the event, or simply absent or elsewhere. There is no indefinite, which is the very meaning of the margin. The world of the digital is hermetic such that it exerts a rigid and absolute control over its domain.

Ruled Bodies

Think about the contingency of human difference. Everyone has a characteristic way of moving her body, personal patterns of motion at both refined and gross scales, each pattern unique. That uniqueness is not determined algorithmically. It is not a matter of a set of variable factors. It cannot be captured by rules for two reasons: First, it is not an isolable behavior, but engaged with a world, with a body in all of its parts and all of their details. Second, it develops according to literally countless criteria, all the accidents by which an infant (even a fetus) trains its body, its nerves, finding paths of motion that suit its particular bodily relations. Those patterns, like they say about language, include endless variation, accident upon accident, or in a word: contingency.

One could simulate this with a powerful computer: large numbers of simulated people with simulated embodiment, each body unique by virtue of a statistical variation across many variables that describe bodies and their development. Imagine millions or billions of data points for each body. One could simulate the growth of these bodies from infancy, programming each to learn its body by conjugating its particular physical characteristics with a unique set of stimuli in the world, responsive to a detailed simulated physics, a highly local simulated materiality, a cutting edge kinesiology. How would this differ experientially from the real world? Doesn't the inconceivable complexity of this simulation come to equal, for all practical purposes, the real?

Even with a complexity that we cannot measure to be less than that of the actual, even if the digital included the resolution of the real, Planck's interval perhaps, even then this simulation would be missing something, would narrow its range of possibility due to a lack of contingency, a ruliness, the important difference from the real. It is the *production* of difference, constitutive of the real, that the digital can never accomplish. In the digital, all production is ruled. The motion of each person would be a calculated set of variables, a specifiable set of values, and however many values there are, they have already determined all that counts as the unique motion of a given individual. But, as a set of variables, they have also determined what counts as motion for every individual. They constitute a rule that ties variables to a model.

Maybe there are ways to unhinge a rule within the digital. What if the digital simulated not only the individual evolution of bodily motion but also the formation of pathways that tie that motion to all sorts of other stimuli, from affective states, to Oedipal relations, to memories recent and ancient? After all, our actual bodies learn their habits not just according to models of physics and kinesiology, but also in relation to all the events that touch us from infancy.

Such simulation sounds complicated, but we are asking, at least for the moment, about digital ontology, what the digital can do structurally, not practically. Surely simulating a whole nervous system as it develops would escape the rules that govern a simulated unique body under unique stimuli in a highly detailed model of physics. But that simulation of the nervous system would itself operate according to a rule, a rule that decides what a neural pathway is, a rule that decides what parts of the simulated world a pathway can connect to, a rule that decides how the set of variables that describe the characteristic motion of a body are articulated in relation to that nervous system. The rule allows for all sorts of influence, indescribably rich possibilities, but as a rule it remains inviolable, not subject to contingency, a world of necessity.

If learning is a habituation, then it also shapes the learner: a person who learns to dance moves differently thereafter. One who assembles bicycles comes to see the world with altered potential, finding new paths, a different space from the driver or walker. Like most learning, learning to play the cello requires many different kinds of training. You certainly must train your body: how to sit, how to see, how to listen, and of course how to move the bow, how precisely to depress the strings against the fingerboard. Subtly, the cello training also alters your relationship to the rest of the world, even when you aren't playing. You hear differently, you pay attention differently, your body grows a supplemental vocabulary, a different receptivity. You have been *celloed,* refashioned by a learning.

Almost everything requires training. Infants have to learn how to use their limbs. A newborn refines her suckling instinct, figuring out what works and developing her technique with its subtle signature, the way she holds her head, the placement of limbs and body, all worked out in an accord, an accommodation with the breast or the bottle and the body attached to it, which of course brings its own rhythms, its own forms, that also shape the infant's learning. The twenty-two-day-old human embryo heart practices beating even prior to the involvement of the nervous system, training its individual cells to coordinate their contractions. We refine and specialize our abilities, training hearts, lungs, brains, and limbs for different efficiencies, athletic conditioning, yogic concentration, bursts of effort to maximize force, or steady output for a long haul. The point is that every training changes the one who is trained. It isn't just a matter of acquiring a new skill, but of becoming a new person, again and again.

Digital training, or training in relation to the digital, often constitutes a special case of learning, as there are only hard edges, getting it right or getting it wrong. Training in the actual involves a margin of subtlety, and expertise often means gaining a mastery over this margin, finding ways to finesse those tiny differences that may exert an outsized influence on the outcome. The digital includes no margin, as its discrete objects and actions are perfectly definite. Part of what one learns in the digital is its absolutism, its all-or-nothing character. Make the right move and win the game or save the file or filter the image; make the wrong move and lose, or crash the machine, or simply fail to have any effect at all. It isn't about creative invention through nuance; it's about orthodoxy, correctness, *yes* or *no*.

In the actual, one often learns not principally by rote, but by intuition, by guesswork, the application of an extended sensibility. Even examining an unfamiliar situation or an object never before seen, one knows that this situation or object exhibits a sense, an integrity that makes it what it is, but also (and just as essentially or basically) maintains ongoing and prehistoric relations with the rest of the world. The assertion of a thing's integral being, its being that thing, is also an assertion of its worldly origins, its having come from somewhere, its possibilities of resting or moving, growing or decaying, combining or dissolving. The world has a way of being that often remains mysterious or challenging until it is revealed, but the possibilities of revelation are sensible, if unbounded. To learn in such a case is to sense beyond the positive immediacy of the situation, to have an idea of what comes next, of what one does not yet know, according to a logic of sense that grows in each of us as we grow in the world. Not so in the digital case, where the logical reduction of everything places it in a world beyond intuition, a world that can be grasped only intellectually, through reason or logic. Intuitive grasp can be simulated, but the edges of that simulation will fray, revealing its isolation, its hermetic confinement to its own digital domain.

As we have seen, digital devices admit certain kinds of interaction, a delimited vocabulary of possible inputs. Computers that can process language generally require particular kinds of language, only responding (appropriately) to some linguistic input, specific domains of language. Computers that accept gestures accept certain kinds of gesture, maybe mouse moves and clicks, and keystrokes, but holding up a palm will not stop the computer from proceeding.[1] As with any tool, one must learn to use a computer, which means learning its affordances, learning its way of accepting input. But it is characteristic of a tool, like a computer, that it serves a purpose, has an "in-order-to," which establishes fairly clear boundaries between effective interaction and other interaction. One can stare passively at a computer or stroke its case with gentle fingers, but that isn't what it's for and doesn't really address it as a computer.

Jim Campbell

The installation artist Jim Campbell exposes in his art the rigidity and empty formalism of the digital alongside his fascination with digital media. His works often use pointillist light sculptures to demonstrate how points, even en masse, tend to fall short of a plenary significance, as groups of LEDs remain individuated spots of light and do not form an image, at least not without some further intervention outside of the digital to bring them into a relation with each other in the viewer's perception. Typically, Campbell sets those points of light in motion or installs a diffusive layer over the light, and this additional dimension allows the information encoded in the light sculpture to reappear in analog form, available to human apperception.

In a thoughtful piece published in *Leonardo*, Campbell expresses some of the principles driving his work. "Programs are mathematical representations: they have to be defined mathematically. This brings interesting questions to the artistic process when an artist is forced to transform a concept, an emotion or an intuition into a logical representation. This is a difficult thing to do without trivializing the original concept. What often happens during this reductive and transformational process is that the subtlety in the work is lost simply because of the fact that things have to be defined with mathematical precision" (134).

Campbell rues the reduction of the artwork when rendered primarily through digital means, as the subtlety of material experience tends, in the digital, to be replaced by the hard edges and perfect definition of mathematical representation. He notes that, initially, effective simulations of engagement and responsiveness in a digital work take advantage of the viewer's unfamiliarity with the work's operation, which adds a sense of depth or mystery, but that sooner or later the art reveals its determinate pattern, its inability to produce new experience. "The point is that often the first time an interface is experienced it is perceived as being responsive, but if the interface is experienced again it becomes controllable. The second time it is not a question but a command." Following commands is what computers do well; indeed, it is the only thing that computers can do. "The computer industry's goal of making computers and programs smarter is simply to make computers more efficient at being controlled by the user to get a job done. Why should they do anything else? This is generally what we want computers for: we want them to be passive slaves" (133).

Though he uses somewhat different terms, Campbell's central emphasis is that worthy art leverages an essential contingency, so that each encounter with it can bring out something new. The artwork needn't change in any way measurable in positivist metrics; on the contrary, painting and sculpture (and nontemporal art generally) include as part of their meaning a certain stolidity, a stability that can even feel defiant or proud in contrast with the motion and change around them. But even in their fixity, artworks house a depth, an indeterminacy at the margins that guarantees a novel experience at each encounter, something to return to, something that renews itself always again. This subtle edge of renewal is what Campbell discovers to be lacking in the digital, which fixes its objects precisely through mathematical representation; in the digital, even an object in motion, even the introjection of randomness, still produces a fixed pattern or illuminates an ultimate constraint that eventually ceases to offer anything new.

He addresses directly the inadequacy of the merely random as an ersatz contingency, noting that randomness fails to make sense, while contingency calls forth the spontaneity of reason: "The ability of a program to make a truly arbitrary decision, an unmotivated decision, is often used to model many naturally occurring processes, but usually it is an inaccurate model. Typically the only characteristic in common with the process being modeled is unpredictability. Irrational behavior, for example, is unpredictable, but it is anything but arbitrary. If many irrational interactions occur within a communication, and the actions point to the same set of hidden motivating forces, they will begin to reveal what these motivations are. [By contrast, a] number of random actions will point in many different directions, creating nothing but confusion" (135).

Digital devices are not especially forgiving or flexible. On the contrary, they could aptly be called brittle; they have rules, and are practically made of those rules. And whatever does not respect the rules cannot enter into the machine, cannot figure as part of the machine's calculated world. Those rules extend to and govern even the interface, the surface of encounter between the device and the user. What the machine offers to the user is the outcome of rule-based calculations, and what the user can offer to the machine is also circumscribed by a set of rules. Stated informally, this means that, if you want the machine to do something, you have to address it in a language it can understand. More than that, you have to anticipate and respond to its rules, mirror its dual logics (of representation and binarity), confine yourself to its input modalities. In brief, you must think and act like a digital device. (Think about liking your friend at school or work versus "liking" something or "friending" someone on Facebook.)

Friedman's analysis hints at the threat of dehumanization associated with thinking like a machine, but also celebrates that altered consciousness as a primary source of the pleasure of digital interaction. When one replaces one's cognitive process with "the blank hum of computation," there is a kind of satisfaction of becoming automated, borrowing some of the efficiency and unflinching certainty from the logic of the machine, a circuit of logic into which the user is inserted. Friedman describes the experience of playing certain digital games, but it is not only gaming that generates the distinctive affect Friedman identifies with digital interaction. That blank hum of computation could also be described as the feeling of becoming-necessary, inhabiting a world of necessity. Human cognition ordinarily (and almost always) confronts the uncertainty or contingency of decision-making, the peril that hangs over every living moment, but to cede cognition to the digital machine offers an escape from that precarity, a context in which one need no longer worry about anything, but simply proceed with each next step, treating it as a necessary gesture in the circuit of interaction that flows through user and machine. This is surely one of the poles of libidinal investment in our digital devices, relinquishing the anxiety of responsible subjectivity to insert oneself into a circuit that establishes *and* satisfies one's desire. (The digital is the lover that never disappoints.)

In another essay on digital gaming from his 2006 collection *Gaming,* Alexander Galloway cites an additional if ambivalent consequence of thinking like a machine. He underlines the strong parallels between machinic action and the logics of daily life in the twenty-first century. To live in American culture today is to negotiate a nearly endless gauntlet of informatic structures, choices, categories, formal characteristics, logical deductions, and so on. Building on Gilles Deleuze's analysis in "Postscript on the Societies of Control," Galloway describes contemporary culture, at least in those parts of the world most dominated by the digital, as built around flexible networks, deconstructible entities that break down into parts, with a broad emphasis on information as opposed to material, completing a transition from the industrial to the digital age. Though Galloway confines his claim to the act of digital gaming, his point, like Friedman's, applies to computing more broadly: to learn a digital game or a piece of software is to practice a way of thinking and acting "coterminous" with political control in today's world (92). As the digital has spread into every corner of the world and every domain of human endeavor, the world confronts its inhabitants informatically, so that our interface to the quotidian reality in which we live increasingly resembles a digital interface and demands the same sorts of interaction for effective control. Digital games often make their informatic or algorithmic design explicit, such that playing the game means quite obviously figuring out how an algorithm works, and Galloway marvels at this seemingly brazen embrace of the paradoxical equivocation of play *as* rule.

The mirror in which the digital and the world are reflected in each other, polished by the Enlightenment ideology that spurs the rise of digital technology, benefits greatly from that foremost feature of the digital, its ruliness. As its rules restrict a digital technology to certain possibilities of input and output, so its users have no choice but to abide by those same rules as they use the machine. There is little room for play, and more than other media, the digital establishes guidelines for how it should be engaged. It offers pathways that are pregiven, ways of use that are standardized, prescribed, and conventional. To figure out how to use a piece of software is a matter of discovery and not invention; one must follow the intended path, for no other path is likely productive. Digital training is therefore unusually rigid, tightly enforcing its injunction to think like a machine. And coming to think like a machine, especially when that habit of thought yields precisely the desired results, one is the more likely to continue to think that way even away from the machine.

Computer Dialog

Often lacking the resources (or the motivation) to include the most robust Natural Language Processing (NLP), digital games find other ways to simulate conversation between the player (who controls the avatar) and the game (or some character in the game that is not the player's avatar). Text-based games, which saw their retail heyday in the 1980s, when Infocom released many such games, require the player to enter short phrases of typed text. Many text-based games allowed only two-word phrases of input, though some could parse longer sentences. (As personal computing devices have become more powerful and NLP software more advanced, more modern text-based games are often capable of parsing much more complicated inputs.) These 1980s games were most often designed around puzzles to be solved by the player, who not only had to figure out how to open locked doors or get across deep ravines, but also had to divine which two-word phrases the game could actually understand. "Cross ravine" might prompt a response from the game like, "I do not know how to do that to the ravine," or even "I don't see a cross here," whereas "swing rope," though semantically underdetermined, might be just what the game was waiting for: "You find yourself on the other side of the ravine."

A popular technique in more recent games is to give the player the option to choose from among various things to say when it is the player's turn to speak in the simulated conversation. This can be engaging and can encourage the player to assume a role within the game or maybe to choose the option that seems most like something she would say in real life (were the real life context similar to the game context at that junction). But it tends also to feel uncomfortably constraining, and even when the game limits the amount of time for the player to select her conversational gambit, this back-and-forth is less like the precarious spontaneity of actual conversation and more like a strategic selection to achieve a best possible outcome.

Conversation in the actual world with a real person is essentially informed by the possibility, hovering at the edges of the discourse, of saying something wholly unexpected. Making sense is spontaneous, creative, unbounded. And even when one sticks to conventional conversational gestures, these are modulated by one's singular self in that singular moment. That is, it is not just the linguistic content that one utters that evinces the contingency of conversation, but also the nuanced inflection, the tone, the pace, the pitch, the accent, the diction, and all of the subtle relations among these that produce one's meaning, which is therefore always a production, always catalyzed by the unique and plenary context of the instant.

Which is why even a more sophisticated digital-language processor cannot finally get conversation correct: it necessarily employs a model, a model that might generate and even take account of more and more nuance but that can never achieve the unbounded spontaneity of persons. Any model will have already decided how it hears an utterance, what counts as the elements of meaning and how their relations carry that meaning. Nothing outside of these parameters can make meaning to the machine.

The crux of the argument is therefore that the production of meaning, the way the world makes sense, is spontaneous and irreducible. We cannot delimit the elements of meaning, nor can we offer a final frame on the process of making meaning. Even as meaning-making follows many patterns analyzable and reproducible in complicated rule sets, those patterns are only the form of meaning and not its content. The spontaneity of sense, like the necessity of contingency, is not a rare or occasional intrusion on an otherwise predictable and unperturbed production of meaning; rather, the spontaneity of meaning is the element of indeterminacy out of which sense can arise. It is only in a context that includes the possibility of the radically unlikely, the unconceived, that one can produce sense.

As computers don't think, to think like a machine must be understood as an injunction to highlight and commend the same values of positivism, rationalism, and instrumentalism that ushered in the rise of digital technology and that are themselves promoted by that technology. More specifically, effective use of a digital device means figuring out its predetermined responses, analyzing (if implicitly) the logic of representational structure, then modulating one's desires and constructing one's technique around these positive logics. Just as the computer iteratively breaks down a task into smaller tasks until one arrives at individual commands (which are themselves further broken down into machine commands and then logic gates), so one must break down one's digital task, arriving by individual steps at one's rationally realizable goal through an instrumentalization of the digital mechanism. No doubt it is at least possible to *play around* at the computer, to try this or that, to mash buttons and see what happens, to select each menu item to see what it does, but this play quickly reveals that there isn't really much play in the machine, as each capricious gesture produces a definitive result that follows the deterministic model that governs the entire machine as a rigid logic, like well-trained soldiers proudly snapping to attention at the command of their drunken captain. The machine demonstrates forcefully its inescapable rationalism, and so also foregrounds its available instrumentality. It begs to be commanded and promises to do precisely what is asked of it.

Is it true that the computer doesn't think? What about all the artificial intelligence, the deep learning, the expert systems? Those very labels attempt to invoke the act of thinking without naming it as such. It is not important whether one extends the attribute of *thought* to digital technologies—after all, it isn't clear what it means to say that humans think if it isn't just an affirmation of certain recognizable behaviors that might be simulated on a machine— still digital thought shows a real distinction from human thought on the basis of spontaneity. To say that reason is spontaneous is to say that it is never finally circumscribed by a set of rules but always might appeal to a further rule beyond the given set. Reason invents itself each time, even when it appears to think the same thing or make the same move. Spontaneity doesn't mean that the application of a familiar rule (or set of rules) is necessarily an instance of unreason, but it does mean that that application of rules has been selected spontaneously, in the face of the availability of an appeal beyond those rules. Even when one makes the same judgment in two similar cases, each judgment remains spontaneous because the consistency is preserved as a genuine choice of reason and not simply a limitation or default option. In other words, thought, too, is constituted by its spontaneity; to think is to think anew.

Can't one approach a computer as one might any instrument, from a pencil to a saxophone, with no particular idea of where one is going? With a pencil, one can let one's hand wander across the page, perhaps applying greater intention once the lines start to suggest a direction (a pattern, a figure, a method) to the sketcher. With a saxophone, a trained musician can make sounds without aim by letting fingers and embouchure and breath go their own way, almost as though someone else were playing, and again the musician might exert a firmer will as she aims to reproduce something she just heard herself play or complete an accidental melody on a satisfying cadence. In both cases, habit likely takes over where intent steps aside, such that it is surprisingly difficult to make contact with the unfamiliar, the outside of one's practice.

It seems more difficult to see how this *accidental* or *automatic art* might work at a computer, for habit in the digital is rigorously enforced, each step engenders a predictable outcome, and a task can be accomplished only in steps. It would be like forcing the accidental sketch artist to choose, consciously, at each stroke, which direction and how much pressure to apply. With the right computer interface, one might let one's hand guide the mouse with but a minimal exercise of will, producing corresponding lines on the screen, when working in a drawing program. But even then, the tight connection of necessity, the rule that ties the hand to the onscreen result, will cleanse the production of the aura of indeterminacy that surrounds real-world activities; those digitally created lines will be less free, less open to the will of the universe, and more determined by the will of the user, however much denied. Perhaps this has to do with the mediated nature of the drawn lines: the unwilled motion of the hand (as on a Ouija board[2]) is made more determinate by virtue of the deterministic program that reads the mouse's motion and uses it to determine the pixels to connect on the screen.

A software instantiation of John Conway's 1970 Game of Life (a famous and simple example of a system of cellular automata) seems to allow a degree of will-less or random input, as one can sweep the mouse (or other input device) across the matrix of cells, randomly populating some of them, and the results of the iterative application of the system rules, starting from the unwilled initial situation, are often surprising, or at least surprisingly complex, until the system invariably settles into recognizable patterns or stasis. But this is a fairly limited "automatic" creativity, as the behavior of the software varies within a small range of possibilities: each cell can turn on or off at each iteration. Then again, turning on and off individual cells is precisely what every digital device does as its essential process, so one should not understate the potential complexity of Game of Life.[3]

Complexity, however, is not really the principal strength of the digital. Instead, it excels at numerical magnitude, the consideration and comparison of mind-boggling numbers of cases. Arguably, number-crunching deals with a particular variety of complexity, wherein any given datum is comprehensible on its own and it is only the aggregation of overwhelming numbers of data that exceeds a human's capacity to keep track. When we use algorithms to examine big data, the underlying assumption is that similarities or correlations (or other patterns) in the data set tell us something worthwhile about the data and about the relationships between individual data and the whole set. When we use a Big Data–based system, say, to predict which molecule will best attach to a protein at a particular site, we are assuming that the answer to that question is calculable as some sort of rule, however complicated and statistically nuanced. We are assuming that there is a strict logic to how things work (or at least to how this particular thing works), a logic that may not be perceivable to an unaided human observer because of the sheer number of possibilities or because of its high sensitivity to variation (or nonlinearity), but a logic nevertheless. To the extent that digital analysis is successful, we confirm that much of the world does indeed proceed according to such logics, that the world is, in useful ways, ruly.

Which brings the analysis back to the bottom line: its dampened contingency weakens the digital's participation in subtle but essentially human values, but also allows it to capture, reproduce, and manipulate much of the world as we see it. *The digital's restrictions are its affordances.* As we live increasingly in a world shaped by and mediated by digital technologies, we adopt the values or ways of being that it promulgates, carrying them beyond those technologies into our everyday lives, our relationships to self and others, our desires, aims, means, and ends. So that, under the sway of the digital, the differences between our lived worlds and the digital domain seem ever slighter, the contingency that sustains creativity and much else of value comes to feel dispensable, and we experience less and less dissonance as we pass back and forth between simulation and reality. For those who have already endorsed a tacit positivism, this progress of the digital can only be a source of optimism: bring on the world of necessity. For everyone else, we are left wondering whether the loss is worth the gain.

The Digital and the Humanities

Digital Humanities (DH) has been the object of much envy among humanists, along with intense derision. Sometimes celebrated as the salvation of a besieged humanities broken by perpetual crisis, DH also suffers accusations of shallow hipsterism and, even worse, complicity with the neoliberalization of the Academy. It has been in some places relegated to the status of a service without full disciplinary standing, but has elsewhere attracted earmarked institutional funding, affirming its legitimacy as a field with an inherent value recognized beyond disciplinary humanities. There remains little consensus about its definition, which is among its most debated topics, and it has over its history strategically recalibrated its focus (or added new modes of inquiry) to sustain itself in the face of criticism, shifting or growing when challenges start to appear insurmountable. DH seems always beleaguered, but it also appears to have secured its future in the Academy and cemented itself as an essential mode of humanistic research.

The analysis in this book has emphasized, likely to the consternation of many, the rigidity, discreteness, and positivist character of the digital. This might set the digital at odds with the humanities, which traffic less in absolutes and perfect definitions and more in contentious interpretations and connotative analysis, the generation of thought rather than its arrival at a fixed point. Nevertheless, there is no reason to think that the humanities and the digital cannot meet productively. After all, the humanities is the art of interpretation, and is as such prepared to interpret just about anything, including formal languages, logical operations, statistical evidence, and all of the positivist and rigid objects and actions that belong to the digital domain. The computer may not be itself a humanistic interpretation machine, but what it does and what it produces are wholly available for interpretation and critique. And, indeed, this describes the core undertaking of traditional DH: the digital machine performs statistical analyses on sets of data, yielding various summary conclusions or locating anomalies, and then the humanist scholar attempts to *interpret* the significance of those data.

In its traditional guise, DH uses statistical analysis or representational techniques such as visualization or sonification to *expose* facets of large data sets that are not readily apparent to unaided observation. The digital tool, properly employed, can reveal patterns in a set of data that unaided human observers would likely overlook because those patterns are too subtle or too (numerically) complex or because those patterns emerge from ways of looking at the data that human observers do not typically employ. For instance, a data set of sentences may demonstrate a regularity in the way the initial letter of a word follows the terminal letter of the previous word—say, words ending in e are followed unusually frequently by words starting with s—and while a statistical analysis, properly directed, would easily discover such a pattern, a human observer would probably not notice it because we don't tend to read sentences or words by a conspicuous consideration of the letters in adjacent words, except perhaps where there is alliteration.

Without explicit direction, digital technology does not know what to look for in the data, as the example just given illustrates. The DH researcher must decide how to analyze the data or what queries to make on the data set. There are plenty of standard algorithmic analysis methods, and many of those are available in prepackaged software-analysis bundles, making it possible for scholars with a limited expertise regarding statistics or other computational methods to conduct an analysis. The brittleness of the digital means that the data must be properly formatted in order to be submitted to the data-analysis software, but there are also automated tools to check and correct data formatting. Because a researcher must generally know what she is looking for, so that she can direct the computer to search for it, opportunities for genuine serendipity in DH research are restricted, though not wholly absent. But if techniques of text analysis such as *topic modeling* and *principal component analysis* (the former identifies commonalities among the members of groups [of data points], while the latter finds the most distinctive elements that distinguish one group from another), for example, frequently reveal surprising or unexpected aspects of the texts being analyzed, it is nevertheless the case that these techniques have already restricted the domain of discovery by necessarily imposing hard limits regarding which objects can enter into the statistical calculations. To be studied at all in DH, phenomena have to be defined, and defined as something measurable, with either an order or a categorization. Thus, the digital calculus of DH is rule bound, rarely producing genuinely novel knowledge in the computational analysis phase. It follows a logic of discrete possibility rather than open-ended potential, for the questions to be addressed through computation must be formulated in advance. All too often, DH research concludes with these statistical analyses, whereas one would expect the most valuable insights to come only after the computation, insights based on humanistic interpretation of the data.

Then again, some proponents of DH celebrate it precisely as an opportunity for discovery. The method of deformation, as championed for example in Stephen Ramsay's 2011 *Reading Machines*, proposes algorithmic alterations of texts (or parts of texts) that proceed in a deinstrumentalized fashion, where the researcher has not decided in advance what she hopes to demonstrate. Ramsay suggests that, by executing certain rule-based methods of reordering, supplementing, re-presenting, statistically describing, and in general *deforming* texts, we put ourselves in a position to learn something new about them. The modesty of this proposal is impressive. It does not guarantee any particular results, but rather expressly conceives its method as experimental in the sense that it may or may not prove productive. Moreover, it is not decided in advance of the deformation what conclusions a humanist researcher might draw on the basis of the deformed text. The computationally aided experiment functions as fodder for further interpretive steps that do not presuppose any particular meaning that the deformation might reveal, steps to be undertaken by the researcher and not the digital machine. In effect, the deformation offers a previously unseen angle on the text that serves as potential inspiration for new insights. This method is thus also admirable for putting the sign in the window, making its (unrestricted) aims and its method explicit: the inclusion of computational tools is not so much about showing something deep, but rather about reorganizing our perceptions of what is already there on the surface so that we might see something new, ready for interpretation. The digital does not generate new knowledge, but creates a perspicuous context in which humanists can discover new insights.

Ramsay's general approach, wherein the computer is used to produce a viewpoint that might jog new ideas, ought to be the core method of responsible DH research. But in practice many factors combine to disguise or weaken this sense of responsibility, such that much DH work stops short of the (humanist) interpretive step, resting content with the (digital) act of analyzing the data. Many projects downplay even the data analysis and simply collect the data to be analyzed, and DH often validates such abortive projects by labeling them *archives*. Certainly it can be immensely valuable to gather data in a digital form and make it available for analysis. But, granting that the discovery or creation of an archive is sometimes an essential step in humanities scholarship, establishing an archive does not on its own generate the critically examined concepts or understanding that are the foremost products of humanist research. DH research is plagued by these proof-of-concept projects, which is understandable because it is usually much more challenging and riskier to endeavor to discover worthwhile insights about analyzed data than to make a digital archive. (Archive creation has plenty of challenges of its own, but they are not mostly the same as the demands that productive interpretation requires.) Though it is a cynical observation, every subfield in and out of the humanities generates only a bit of outstanding original work and a lot of mediocre or deeply flawed work, and in DH, the mediocrity most often takes the form of projects that don't even pretend to insight.

The problem is not that computationally intensive analysis somehow yields results stripped of their humanity by the mathematized algorithms of DH. As mentioned above, there is no reason that a clever and informed humanist cannot generate interpretive insights from any objects of study, including the numerical or categorical results of statistical analysis. The problem is rather that, in a quest for legitimacy in the face of a suspicious and sometimes hostile traditional humanities, DH has self-consciously attempted to emphasize its disciplinary distinction, to claim that it represents a new approach to humanistic research that is also continuous with traditional humanities. In other words, *the problem is that DH claims to be a field.*

But DH is not a field of study. It specifies no distinct object of study, and its distinguishing feature is not an intellectual, but a technical criterion. *DH is humanistic research in which intensive computation plays a significant role.* One might equally grant field status to "legal pad" humanities, or maybe "sitting-down" social sciences. Presumably these parodies of research fields would be disqualified, and DH validated, because DH is distinguished not just by the formal criterion of the use of computation, but by its novel methods and otherwise unobtainable results. That is, the claim is that there is something intellectually distinctive about the sorts of insights to be gained through the use of computationally intensive techniques, that it is a different kind of knowledge. Ideally, DH researchers pursue questions about meaning and value similar to the questions asked in traditional humanities and DH puts those questions to (many of) the same objects that are studied in traditional humanities, but because DH methods yield information about the objects mostly unavailable to traditional humanist researchers, new insights are uniquely possible.

This turns out to be a difficult ideal to uphold. A set of textual data can (unproblematically, though possibly laboriously) be subjected to a topic-modeling analysis or a principal-component analysis. Those operations, which are at bottom statistical, numerical calculations, require only that the data be properly formatted and available. But there is no guarantee that the results will suggest any meaningful conclusions. The specialized training in DH is primarily technical training: how to use the available portfolio of tools and techniques; how to gather, "code," and disseminate data; how to design a flexible, useful interface. Though these are potentially valuable skills, their value in a humanities context lies in their prefatory nature, as they yield perspectives on data that can then be humanistically interpreted. Armed with the techniques of DH, dedicated DH researchers search for objects to which they might apply those techniques, which reverses the research procedure of traditional humanities. A traditional humanist chooses objects of study because those objects are somehow curious, opaque, confusing, controversial, or otherwise wanting an examination. Then she studies the objects, developing questions to ask and applying techniques (if appropriate) suggested by those questions.

By contrast, a DH researcher starts with techniques and goes in search of things to apply them to, often foregoing altogether the humanistic, critical understanding. It is too often forgotten or simply ignored, especially after sometimes laborious efforts to prepare data and treat it computationally, that the processed data generated by DH analyses play the role of evidence and not conclusion—it must then be determined what, if anything, those data are evidence for. Data have a particular epistemological role: they confirm, they disconfirm, or they have nothing to contribute. Because they are data, they don't say anything yet.

In principle, DH research constructs original but not wholly alien objects to study, objects represented by data derived from computational techniques, and those data are then subject to the same sorts of interpretive analysis performed in traditional humanities. But in practice, insisting on DH as a field has resulted in a glut of young researchers whose distinctive expertise lies in the use of computational tools, such that the methodology of DH serves as their claim to institutional legitimacy; insight, formerly the vaunted prize of humanities, is thereafter assigned a secondary and deprecated status. Rather than encouraging its practitioners to reassert the distinctive gesture of interpretation, DH, to maintain its authority in the face of accusations that it lacks a critical emphasis, began to valorize the employment of its tools and the creation of data archives for their own sake, declaring these to be valuable humanistic practices in themselves. Armed with hammers, DH researchers proceed by searching for stray nails.

This is not to suggest that the processes of traditional humanities guarantee insight while DH necessarily falls short of it. Rather, it is a criticism of DH as having all too often, in its enthusiasm for its promising and novel approach, misrecognized its own greatest potential. To restore interpretation as the central practice of DH is to push it into a confrontation with contingency. Whereas computational analysis yields regularities and also indicates outliers in relation to those regularities, the further step of interpretation would demand not the exposition of a regularity or even the identification of exceptional cases, but the kind of insight that arrives through a studied intuition and an openness to accident, when thought arrives in a flash. Humanities, and perhaps all the sciences in which we engage, are domains of questions more than of answers. The computer has yet to ask even one.

‹ 7 ›

But . . .

But don't computers output a fire hose of streaming creativity all the time, contingency be damned? Don't digital machines make it possible to do all sorts of things we just couldn't do without them? Don't they solve problems of engineering and mathematics that could never be overcome by an unaided human person? Don't they discover patterns, generate new knowledge, yield valuable insights that provide new perspectives on the world and the people in it? To watch a machine render the beautiful complexity of a fractal image on the screen, to hear the occasional soulfulness of algorithmically generated music, to interact with a chatbot that appears to anticipate your thoughts, to witness the advances made possible by digital machines in medicine, commerce, travel, sports, communications, design, media, entertainment, and just about every area of human endeavor—isn't this to be awestruck, astonished, and delighted, to encounter what was only recently *unimaginable*? Faced with a digital technology that is practically synonymous with progress, the source of innovation upon innovation, the bearer of one surprise after another, how can it be claimed that the digital lacks creativity, that it plods along doing the same old thing, that it is capable of no more than to rearrange the obvious? Even more implausible, how can this book suggest that the digital dampens *human* creativity, holds *us* back from invention, confines us to familiar and moribund patterns of existence and thought?

GPT-3, the third iteration of a Generative Pre-trained Transformer system developed by OpenAI, is a database in the form of a neural network. Recall that digital, "artificial" neural networks are not programmed with explicit rules, but are instead trained, using positive and negative reinforcement, so as to develop their own rules. Among the many varieties of neural networks, *transformer* networks (which break down a large input into smaller parts and then calculate how closely related each part is to each other part) currently occupy the leading edge of research in natural language generation. And so this commentary on GPT-3 is relevant, *mutatis mutandis,* to other natural language processing (NLP) transformer networks, such as ChatGPT or BERT, which has numerous real-world applications and even helps the Google search engine to make sense of user queries.

GPT-3 was trained by feeding it a huge amount of mostly ordinary text, hundreds of times more text than had ever before been used to train a language-oriented neural network. About eighty percent of the training data were pulled from ordinary websites and other text available on the internet at large, though the system was also fed a good chunk of Wikipedia (over two billion words worth), a significantly bigger chunk of upvoted Reddit posts, and around sixty billion words of published books. Though the details are technical, one could roughly describe all that textual training data as providing *examples* for the system to follow. With the training complete, one can input to the system a few words or a sentence, and GPT-3 will use what it has "learned" from all the examples to produce some more text that it calculates to be a likely sensible continuation of the words or sentences that have been input. Ask it a question, and it will generate an (often) plausible answer. Give it a few sentences of context plus a few words that might start a sentence, and it will complete the sentence and develop one or more full paragraphs, consistent with the supplied context, to make a short story or a journalistic capsule report.

Recognizing the significant potential for mischief (and worse) presented by such a powerful language processing system, OpenAI has restricted access to GPT-3, requiring that scholars and other artificial intelligentsia state their intentions and provide some additional details before they can explore the system. Here is an apt example of GPT-3 at work. The initial prompt provided to the system is in bold. The remaining text was generated by GPT-3.

This book is not a critique of one or another digital technology but a critique of the digital in general based on its underlying technological principles. It is also not a book about politics or the social. Rather it is a phenomenological account of a great lack in every digital technology, the lack of contingency. My main point is that the digital is an ontology that, far from "getting rid" of the object or the subject, fully actualizes them into what they are. This is why the digital is so powerful, but also why it is so dangerous. The digital is so powerful because it actualizes digital things to be exactly what they are. But it is also so dangerous because its ontology is one of perfect control and predictability. For this reason, the digital is one of the most antihuman technologies ever produced, and it will engage us—individually and collectively—in a grand struggle for the soul of technology.

To conduct this critique, I will be providing a thorough account of the ontological status of the digital. First, I will show that the digital is an ontology that fills in the blanks in the real. It is an ontology that is not opposed to the real, but that instead is the ultimate actualization of the real. Second, I will show that the digital is both an ontology of the object and an ontology of the subject. This means that the digital is inescapable, in that it envelops all of existence in its ontology. Before I can show this, however, I will have to examine the basic structure of digital technology from a phenomenological point of view. In the final chapter, I will show how the lack of contingency in the digital is the result of a basic logical or structural dimension of technology that will be called calculative reason.

On the other hand, here is a second example in which the same prompt led GPT-3 to offer a different completion.

This book is not a critique of one or another digital technology but a critique of the digital in general based on its underlying technological principles. It is also not a book about politics or the social. Rather it is a phenomenological account of a great lack in every digital technology, the lack of contingency. The idea of the digital is based on the idea of the digital, or rather the idea of the digital is based on the idea of the digital. This self-referentiality is the core principle of the digital, and although it appears at first glance to be paradoxical (for how can something be based on its own principle?) it is nonetheless the case that the principle of the digital is self-referentiality. The idea of the digital is based on the idea of the digital because the

> idea of the digital is based on the idea of the digital because the idea of the digital is based on the idea of the digital. From here I proceed to the first important conclusion: the idea of the digital is based on the idea of the digital because the idea of the digital is based on the idea of the digital because the idea of the digital is based on the idea of the digital because the idea of the digital is based on the . . .

Though GPT-3 is demonstrably imperfect, it is also astounding in its capacities. Not only can it generate paragraphs of coherent, compelling, and original prose on almost any topic, but it can write computer programs, perform basic reasoning functions, emulate the style of a known author, construct sophisticated poetry in many poetic forms, and create web-formatted tables and charts, each prompted by only a few words of ordinary English. Previous systems were custom-built to perform one or another of these tasks, but GPT-3 can do them all, and it can do all of these things without any special training beyond exposure to that massive trove of text. Which suggests, or maybe even proves, that basic reasoning, prose style, argument, and more are all built in to the way we ordinarily use (written) language.

As a neural network, GPT-3 has never been given explicit rules about how to respond to a query or how language generally works. It has not been told, for instance, what an adjective is or that verbs must be conjugated to match the subject of a sentence. If it nevertheless successfully conjugates verbs and uses adjectives appropriately when generating its uncannily human-like prose, that is because it stores the conjugation rules and the most appropriate adjectives for a given context in its probability matrix. Neural networks are made of a bunch of interconnected artificial neurons, which, something like bits, are generally in one of two states, sending or not sending. When a neuron receives a signal, it has a certain probability, refined through training, of passing that signal along to the neurons to which it is connected. In the case of GPT-3, there are 175 billion such probabilities, also called "parameters." Thus, a neural network like GPT-3 does not have explicit rules for computation, but its massive network of probabilities encodes rules immanently, as probabilistic pathways through its web of neurons.

In effect, GPT-3 archives in those parameters relationships among words, particularly relationships of proximity. Its mode of operation is to accept a string of words as input and then to calculate, based on its probabilistic matrix of word proximities, which word (or words) make most sense as what comes next.

To put it otherwise, the one thing GPT-3 *knows* is which words go together well. (Shades of *Michelle, ma belle!*) But there is no knower behind the knowing: the prodigious linguistic abilities it demonstrates are the outcomes of statistical and probabilistic calculations. They don't reflect any situated knowledge or experience of the world. One can legitimately say that GPT-3 has a grasp of grammar or understands arithmetic inasmuch as one is affirming thereby no more than its demonstrated ability to offer appropriate responses to queries that, when posed to a human, would call on such understandings. Then again, perhaps human understanding is also nothing more than a demonstrable ability. Is understanding only a matter of passing a Turing test, the ability to respond in a way that simulates human being so well as to be generally undistinguishable from it?

Will Douglas Heaven, writing in *MIT Technology Review* in 2020, echoes many other commentators when he writes that GPT-3 is extremely impressive but also "mindless": "GPT-3's human-like output and striking versatility are the results of excellent engineering, not genuine smarts." And his quote from OpenAI cofounder Sam Altman expresses the consensus bottom line: "AI is going to change the world, but GPT-3 is just a very early glimpse. We have a lot still to figure out." This promise of a better digital future feels somewhat hackneyed after half a century of breathless assurances from tech gurus and tech profit centers, but in this instance, it also reflects an admirable humility motivated by the manifest weaknesses of GPT-3. Its generated stories frequently devolve, as above, into the endless repetition of one or two sentences. It fails at some first-grade-level tasks of simple spatial or logical reasoning. It can offer obviously incorrect answers to problems of basic arithmetic, particularly when working with numbers of more than a few digits. It shows no particular regard for truth, as its reports feature fabricated quotations and confabulated factoids. This indifference to reality is hardly surprising, as statistical relationships among words do not also comprehend the real to which they ostensibly refer, and the training corpus includes fiction, propaganda, and outright lies, which have no special status among the many texts of valid facts and authentic states of affairs that are also used to train GPT-3. All of it is just strings of characters, data for processing and pattern recognition.

As with all digital technology, GPT-3's behavior, both impressive and ridiculous, comes down to a manipulation of form. It treats language as statistically significant patterns of letters. The elemental unit for GPT-3 is not the word, but the *token,* one of a prescribed set of common sequences of letters (including punctuation, diacritical marks, and other symbols) found in written language. Tokens are, on average, about four letters long or about three-quarters of a word, though some longer words are also single tokens. During training, GPT-3 has to "figure out" the existence of words as one of the statistical patterns that characterize sequences of tokens, by noting the inclusion of a space or punctuation mark every couple of tokens. If it captures context, if it captures anything, it's because that thing can be captured as a statistical relationship among tokens. So, what GPT-3 finally demonstrates is that detailed statistical relationships among tokens are quite telling. Those relationships, archived as 175 billion structurally sequenced probabilities, can simulate much of language, providing a serviceable instrumental access to language generation. (GPT-3 suggested, in response to a prompt about its own limitations, that it could, for example, be helpful as an aid to translators; it could do a rough translation quickly, it said, allowing the translator to focus on the more challenging and nuanced aspects of the translation.)

Does this simulated language generation, confined to formal analysis, somehow transcend its formality to bring contingency into play? There are symptoms in the behavior of the system that we might associate with contingency. GPT-3 is a stochastic model, meaning that it generates *likely* responses, where likelihood is measured not absolutely but probabilistically. The inclusion of elements of randomness (really *pseudorandomness*; see the "Slot Machines" vignette in chapter 4) in its calculations seems to soften its determinacy and bring it closer to something like contingency.

Randomness is not contingency, but when meted out judiciously, randomness can provide a simulacrum of contingency for a while, at least until it begins to reveal its chaotic senselessness.

Looking more closely, we might say that the model offers a simulation of contingency inasmuch as all 175 billion parameters are determining factors in any given response, including parameters that become negligible to the point of irrelevance for a given calculation. Instead of the inviolable rules of explicit programming, the neural network embeds its capacities in an immanent web of probabilities and associated neurons, such that only the entire web serves all at once as a final rule, and one cannot know whether there are, captured in that network of neurons, any absolute or unmovable limits to language production. The reason encoded in GPT-3 is immanent to it; it is only the system as a whole, the entire neural network, that can be said to cause or explain a given outcome. Just as contingency in the actual is the potential for any factor to play a role, so the simulated contingency of GPT-3 could be understood as the potential involvement of any parameter, even if only some parameters actually bear on a given calculation. In a neural network, everything is touching everything, but everything in this case is the space strictly defined as 175 billion parameters and the artificial neurons whose probabilities they determine.

Perhaps embarrassed by its lapses, many commentators, like the OpenAI cofounder quoted above, emphasize, more than its (often) incredible prose, GPT-3's slightly ominous but ultimately optimistic promise regarding the *future* of AI. It easily outclasses its predecessor, GPT-2, but the only significant differences between the two generations are the size of the model (the number of parameters) and the amount of data used in training. With even more parameters and even more reams of text as training data, will GPT-4 (or 5, or 12) unproblematically pass a Turing test, responding to queries, engaging in conversation, demonstrating a general familiarity with the world, such that human observers cannot tell that it is an automated system, but are fooled into believing it to be another person? Would such a system finally include contingency, inventing not just new responses, but whole new categories of responses, somehow overflowing its formal necessity? Could it mature, evolving over time the way human beings do? Would it be capable of generating language that is not just surprising—for strategically deployed randomness is sufficient to induce surprise—but also somehow consistent? Or will AI, as long as it is built from digital principles, suffer permanently from its lack of contingency, revealing to its users the hard edges that mark the limit of its simulated creativity?

Even if GPT-3 does not escape its confinement to a pure formalism of tokens, its amazing performance demonstrates just how much of the world, or at least of language, can be described in terms of sequences of letters and their statistically measurable relations, a manifest victory for positivism. The model manages to construct fantastic prose by calculating one token at a time, where each next token reflects the exit point of a ball bouncing through a gigantic pachinko machine, with 175 billion brass pins through which the ball caroms and descends, until it comes to rest at a particular token, whereupon the next ball is launched. The intelligence of the system resides in the arrangement and shape of the brass pins, which have each been biased by extensive training to deflect the ball more often one way than another. If humans do not produce language by stringing letters together, we learn nevertheless from GPT-3 just how beholden we are to a world become positive, as its successful operation proves that our language, too, takes many of its cues from a formal apprehension of the world around us. Do our speech and writing reduce to that formalism, or do they always also defy positive formulation? Will it turn out that human language is entirely amenable to calculation?

On the order of the number of neurons in a human brain, 175 billion is a big number, though the artificial neurons of a neural network typically have fewer interconnections than do actual neurons. But 175 billion parameters allows the constitution of plenty of intricate and complex formal relations, enough to capture much of language and even more. Sheer numeracy is one of the principal strategies of the digital, as it can make up for a lot of weaknesses. Throwing more bits at a problem often yields, if not solutions, then adequate workarounds that, from the user's perspective, are more than sufficient. A digital photograph might have some jagged edges when viewed up close, but quadrupling the number of pixels could move the problem, called "aliasing," beyond the capacity of human vision. A set of data about people fails to predict with enough accuracy the voting habits of the individuals in that population, but adding more data, differentiating the population more finely, allows more refined weighting of the data set and a more accurate prediction. Given five desktop-background images, a person likely feels as though she is making a choice; given thousands of images, she may feel as though she is individually tailoring her machine to her unique taste.

The question, as raised already in chapter 5, is whether epistemology or ontology is what really matters, whether an experience that *feels* inventive can trump the *real* lack of contingency in the digital. If conversing with GPT-3 offers rewards on a par with human intercourse, then maybe it does not matter that it is a neural network governed exclusively by formal, positivist principles. If a graphic designer can choose from millions of different placements of a graphic element on a poster, then why should he care that his choices are still finite and that somebody else could, but probably won't, make exactly the same choice? If there are enough filters or colors or buttons, then doesn't the user exercise a true creativity and not just a paint-by-numbers fulfillment of a predestined possibility? If a gamer uses precise timing and a sequence of jumping motions to complete a level of a game without having to fight any of the monsters, then doesn't she feel pleased with herself for discovering this seemingly eclectic strategy, even if, technically, it was a strategy already anticipated in the design of the game and the affordances of her avatar?

So perhaps epistemology just wins out. Maybe if there are enough emojis, with lots of nuanced distinctions among them, they cease to function as iconic conveyors of categorical affect and instead attract to themselves that edge of indeterminacy, that precarious not-yet of significance, that makes something uniquely expressive, productive of meaning. Or, conversely, maybe there never was a peril in the margins that threatens to dissolve meaning, and our language has always been tantamount to a particularly numerous deployment of various emojis. Maybe formal relations among tokens is all that language is made of.

To take choice as equivalent to freedom, to regard what is simply not known as an ersatz for the essentially unknowable, to substitute the disorientation of "too many" for the incomprehensibility of contingency—this is what is required to accept the digital as a domain of legitimate creativity. This book's skepticism about the preference for the ready availability of not-knowing over the difficult pursuit of the unknowable is based on the suspicion, motivated but not proven, that formality will out. The problem is that, for all the differences on offer in a digital environment (all the choices, all the inputs, all the websites, all those tokens laid out in digital space), one is always confronted with the same difference, with the same sort of difference, a difference of discretes, a difference of hard edges, a difference always finally reducible to an integer-subtraction problem. All of the buttons available to the web-page designer, even if there are thousands of options, all rely on a preconception of what a button is, a structure literalized in code that specifies the form of a button, and so also specifies its degrees of variation. The digital is immensely powerful and does indeed offer much that we have never before seen, but a closer look would reveal that even the most novel digital artifact is a variation on a known theme, new wine in an old bottle. The implication is that, after enough interaction, one would finally hit bottom, exhaust the delights of GPT-3, which would reveal all too much method to its madness, a pattern of response that becomes a prose tic and eventually a mindless repetition that expresses only its own mechanism. ("Daisy, Daisy, / Give me your answer, do. . . .")

This consideration also clarifies the gap between M. Beatrice Fazi's conception of computational contingency and the radical contingency that this book situates as the generative force of ontology (see the "Computer Chess" vignette in chapter 4). Acknowledging that each individual step, each logic gate, has a necessary outcome, Fazi brokers a rapprochement between contingency and necessity, locating contingency not in the execution of any one step, but in the entirety of the algorithm. The algorithm's outcome may be necessary, in the sense that there is only one conclusion at which it can arrive, but that one conclusion is nevertheless determined only when the algorithm is actually executed. Before its execution, there is something fundamentally indeterminate, something not-yet, determined only when the algorithm runs, which is therefore *contingent.* This formal or computational contingency is important, says Fazi, and ought to reform our understanding of the digital, an understanding that has never been able to see past the necessity of each step to grasp computational contingency. But this rarefied contingency of computation does not disrupt the digital order, does not tangle up the algorithm with the rest of the universe in a mesh of abundant reason. It preserves the reason by default of digital rationalism, the "rigid, all or none" positivism of the digital, and even leaves intact the instrumentalism that ensures that digital technologies remain so reliably useful (see von Neumann 1963, 303). Even if ontologically contingent, the algorithm's output has already been delimited by its formal structure, if not its particular bit values, a choice among discrete possibilities, not a meshy irruption of the in-principle unknowable.

How does this necessity, on friendly terms with a defanged contingency, inhibit the acts of reading and writing we perform at our digital devices? Surely the text one reads on the screen still solicits the very human act of making meaning, whatever it might or might not mean from the "mindless" machine's perspective. Perhaps this inventive freedom to make meaning in the act of reading (not to mention writing) is where Fazi's computational contingency cashes in its chips. So the question remains, what does the digital contribute to this production of meaning? What constraint or what sort of influence does the regime of necessity exert over the meaning to be made of a digital text or a digital sound or a digital image? What does it matter that a text has been authored in collaboration with a rigid mechanism? What does it matter that the family photo is an encoded set of data, built on a prescribed theory of visuality?

Meaning-making in those instances retains its contact with contingency, reserves a spontaneous productivity that might overthrow any expectation, that might discover or invent a meaning that was not already there. Yet the digital mechanism that produces text, sound, or image tends to lock those digital objects more tightly to their codes, to tie the object by a theory (of text, sound, image, database record) to its intended meaning. The reader is not thereby robbed of inventive freedom, denied a dalliance with contingency, but is more aggressively directed toward an already intended meaning, a canonical act of reading under the authority of the digital ideology. Reading or writing on a digital machine, there is necessarily an introjection of the content of what is read or written onto the binary logic, passing through the representational logic, and these medial logics of the digital inevitably leave their stamp. Freedom *within* constraint is the condition of digital engagement.

One more game, a real one (not hypothetical) this time: Consider the much-hyped video game *Cyberpunk 2077*, developed by the Polish design studio CD Projekt Red. Though its sales numbers notched the "biggest digital game launch of all time," at least according to its Wikipedia page, its many bugs and frequent crashes generated so much bad publicity that a lot of players may have skipped playing it even after buying it, never exploring its fantastically deep and detailed world. Technical faults aside, *Cyberpunk 2077* provides an immense and astoundingly rich environment, including narrative elements, graphical intricacy and variety, a large cast of characters, prolific music and audio cues, and a great deal of attention to detail. Each neighborhood of Night City, the game's setting, has a distinctive architectural style. Every person walking down the sidewalk looks, dresses, and moves a little differently, and many will engage the player-character in brief conversation, perhaps a single reply, revealing different voices and mannerisms. Though side quests (tasks inessential to the core plot but often narratively or thematically connected to it) fall into standardized categories, those categories are quite diverse, and each quest has its own dialog, multiple possible outcomes, and specific locations. There are some games with worlds that feel geographically larger and feature more numerous environments, but there may never yet have been a commercial game with a world as replete with small details, from billboards and graffiti, to individual variation among thousands of weapons and articles of clothing, to a massive library of voice acting (available in multiple languages). The core narrative, comprising a couple of dozen missions, is not in itself a huge undertaking, but there are so many optional side missions that gameplay possibilities are varied and extensive.

Playing through the core missions enacts a tangle of science-fictional plot, involving the cybernetic infection of the player-character's mind and body by another character. The player-character (that is, the player's avatar) is named V, and in the game's early sequences, V is forced to implant a chip into her neuro-circuitry that contains the "engram" of a famous, hard-living, long-dead rock star, Johnny Silverhand. A bullet to V's brain leaves V alive but damages the chip such that it cannot be removed without imperiling V's life. Thereafter, V and Silverhand effectively share the player-character's body, and the plot is driven by V's attempts to have the defective chip removed or neutralized to restore her physical and mental autonomy and to keep Silverhand's engram from taking over her entire nervous system. (The player can choose a gendered body type and a gendered voice type for V. The gender pronouns used herein refer to an avatar with a female body and a feminine voice.) There is plenty of antagonism between V and Silverhand, who compete for control over her body, but they also frequently cooperate or collaborate, and V's attempts to regain sole control over her mind and body eventually dovetail with Silverhand's erstwhile aim from when he was a living person (and not just an engram on a chip), infiltrating and disrupting a major techno-corporate power, thereby twining V and Silverhand in a grudging alliance.

As exemplified in its central plot, the rhetoric of the game purports an ontological ambiguity, muddying many familiar distinctions. Life and death, human and machine, male and female, and as in the main narrative, self and other—these binarities may never have been as clear-cut, as *binary,* as is sometimes imagined (and juridically enforced), but *Cyberpunk 2077* disturbs these distinctions as a recurring motif of the game to generate much of the game's particular significance. One effect of these desegregations is to depict the future as a time of uncertainty and instability, where things refuse easy categorization, a future where one makes choices about how to act or what to do but those choices do not indicate clear outcomes. Morality thereby also becomes muddled in this speculated 2077, and the game frequently locates the player within this moral maelstrom by presenting her with choices in which every option has some undesirable or compromising consequences. Ambiguity complicates moral choice, but at the same time it relieves moral pressure. If consequences are generally unpredictable, if life and death are indistinguishable (and at a basic level they are nearly so in many digital games where death is more of a brief caesura than a final end), then one needn't worry too much about killing.

This dedifferentiation of binary category attempts to introduce a rhetorical complexity often absent in digital games, which typically draw razor-sharp lines between good guys and bad guys, or allies and enemies, and *Cyberpunk 2077* distributes this complexity throughout many dimensions of the game. There are conversations between V and various nonplayer characters (NPCs) that provide essential information and sometimes even advance the plot by giving the player choices among dialog options: different responses can be chosen for V to offer in the course of a conversation with an NPC. In some cases, those choices are consequential, shaping V's relationship to the interlocuting character, and thereby opening or closing possibilities of plot development at later stages of the game. For example, in a given conversation, a friendly response might win the loyalty of the interlocutor, who is then available later in the game as an ally in battle, whereas a neutral or unsympathetic response in the same circumstance might foreclose that alliance or even make an enemy of that NPC. At the end of a mission in which V finds a reproduction of Silverhand's old Porsche, V and Silverhand negotiate their relationship in an extended conversation within V's mind. (To dramatize this internal psychodynamic strife, Silverhand appears here and there in the game as a ghostly hallucinated figure, visible only to V, who can speak with Silverhand subvocally and hear his voice in her mind. The game allows the player to experience this internality as dialog by placing Johnny in a body of his own, acted and voiced by Keanu Reeves, with Johnny's holographic shimmer and greenish tint suggesting a Reeves emerging from the matrix that he mastered in one of his earlier roles.) Depending on which dialog options the player chooses for V's side of that conversation, it is possible to trigger a "secret" ending to the game, an ending that would not have been available had the player chosen different dialog at that point.

Though arguably only an appearance of complexity, this sensitive and subtle relationship between dialog and plot in *Cyberpunk 2077* instigates an unusual characteristic of digital games: by establishing dependencies between specific dialog choices and events in

the narrative future of the game, by concealing this relationship behind unremarkable dialog rather than overtly signaling its import
to the player, and by making the required dialog choices themselves
only subtly distinctive, such that they are not readily recognizable
without foreknowledge, the game is made to feel, at least in this
respect, like a deeply tangled meshwork of interconnected things
and events. The mechanism that connects the dialog choices to the
eventual secret ending feels narratively complex and precarious, at
least against the usual experience of the digital (in games and in
other digital environments) as a narrowly causal structure in which
each choice leads without ambiguity to a predictable and logically
sensible consequence. This game thus represents a world that we
might understand as *contingent,* a world in which even distant
events are subtly interdependent and in which complexity over-
whelms instrumentality. One cannot simply select the secret ending by pressing the "secret ending" button; instead one must meet
the game at its obscure edges, engage not just instrumentally but
affectively, follow intuitions rather than rules, in order to choose
dialog options that might soothe Silverhand's foreign voice in V's
head. The contrast with much other video gaming, including even
other elements of *Cyberpunk 2077,* could not be sharper: so much
gaming involves choosing among options with predictable and
straightforward outcomes and in-game meanings.

Contrasting mightily with the direct causality and stark distinction of most digital gaming, this construction of contingency,
or at least of the appearance of contingency, points to the meaning of the other complexities represented in *Cyberpunk 2077.* The
most ambitious significance of the dissolved binaries in this game
is to challenge, at the level of rhetoric, one of the core limitations
of digital gaming and of digital devices more generally: their discrete and rigid categories, deriving from their underlying binary
operation. Persistently imposing a haze of indistinction around
categories that are typically clear-cut, this game asks players to extend this blurred boundary from the domain of rhetoric to that of
mechanics, to believe that, just as life and death or self and other
are continuous rather than dichotomous, so can a button push or
menu selection engage a fuzzy logic to engender unpredictable
and spontaneous consequences. If games, and digital machines

generally, institute deterministic causal relationships between user action and machine reaction, *Cyberpunk 2077* attempts to persuade its players that things are here more complicated, that the world in which the player acts comprises a messy collection of interdependent elements so numerous and wildly imbricated as to defy human calculation.

But however compelling a player might find this ludic argument, it remains at the level of representation, confined to the game's rhetoric, and it does not transcend the underlying limits of the digital machine, its absolute, deterministic fixity. All the data of the game—all of its states, all of its conditions, all of its NPCs, all of its images and sounds, every line of dialog, every possible action and result, even the secret ending, everything in the game—are encoded in bits, and every bit at any given point in time is either a 0 or a 1. There is no in-between, no ambiguous or indeterminate possible value. Any action by the player, such as the choice of a line of dialog, must be explicitly connected to its consequence in the code that runs the game, revealing a much starker and more rigid causal model underneath the appearance of a contingent world. It is thus only a compelling illusion that Night City is a world of incalculable complexity and interconnectedness. In fact, only the selection of the prescribed three or four lines of dialog will avail the player of the secret ending, a strict causality dressed up as a subtle contingency. There can be no accidental path to that ending, and there can be no ending that is not prewritten into the game's programmed script.

Representation goes a long way. One might enjoy a vertiginous sense of transport into a realm of the indeterminate or ambiguous while playing *Cyberpunk 2077*, persuaded by its rhetorical argument and drawn in by the pleasant disorientation of a digital environment that appears to defy the usual expectations of strict causality and predetermined pathways. This is one of the digital's most ideologically potent techniques, to leverage the power of representation against its digitally discrete ontology, to borrow the seduction of the image as its own, or to wield its unassailable numeracy as a shield that hides its embarrassing finitude. We are only too happy to play along, to enjoy that seduction, to conflate the representation with its medium, to select a line of dialog from a list of choices, because, after all, the player doesn't know what's coming next and can thus easily imagine that she has begotten the NPC's response by engendering it within a field of contingency. It's so satisfying to solve these prescribed puzzles, so impressive to face more choices than one could ever explore, so cool that this year's model has twice as much memory and ten times the number of pixels, that we are unlikely to want to remind ourselves of what's behind the curtain, of the plodding necessity, the step by inevitable step, the yes-or-no logic that props up the whole machine.

If one wishes, and with a bit of extra attention, one can play through *Cyberpunk 2077* without killing any of the NPCs who are the enemies in the game. Sneak up behind an NPC, press a certain key or button, and V will grab that character by the neck. Once thus "grappled," a character can be dispatched with a key press, but notably, one key will direct V to break the NPC's neck, while a different key will instead render the NPC unconscious. And there are weapon add-ons, purchasable from in-game vendors or retrievable from defeated enemies, that modify V's weapons to be nonlethal, so that they too knock out the enemies but do not kill them. Surely this nonlethal option is intended to reduce the perceived brutality and dehumanization, to appease those players who might be uncomfortable with the casual disposal of hundreds of (fictional representations of) human lives over the course of the game.

But if some players might have qualms about the banality of murder in this video game, the game itself does not seem particularly invested in the player's principled stance, for the difference in the game between killing and rendering unconscious is negligible: either way, the incapacitated character ends up prone on the ground, ceases to threaten the player or impede her progress in any way, and so becomes an element of the scenery until the game engine erases the body from the game world. Saving lives by knocking out NPCs makes no difference in the player's accumulated "street cred," nor does it prompt any alteration to the ongoing narrative of the game. There are a couple of points where the player is instructed to undertake a mission to incapacitate but refrain from killing some bad guys, though the game does not then enforce this directive, rewarding the player for completing the mission regardless of whether the enemies are left dead or alive. Online discussion of *Cyberpunk 2077* suggests that the nonlethal option was probably originally a more significant element of the game's design, but may have been abandoned during development, yielding what turns out to be a distinction without a difference.

Though its abortive origins mark it as an anomaly, this almost meaningless distinction between lethal and nonlethal play styles sets in relief the character of digital representation in general. The digital always works through representation, for its images, sounds, texts, and other productions are necessarily the outcomes of operations on encoded numbers. Whatever the digital generates can only be secondary, a re-presentation, as its interfacial artifacts constitute a thin layer built on top of (and strictly produced through) a prior numerical code. And like the signified but inconsequential distinction between lethal and nonlethal in *Cyberpunk 2077*, representations in the digital are given meaning by a deliberate assignment, by the rules that construct each digital representation and specify its relationship to the represented. Such representation thus lacks the contingency that imposes always another meaning in a nondigital representation. Nondigital representation signifies (of course) by virtue of the relationship to the represented, but also accrues meaning precisely inasmuch as it departs from its represented and overflows that relationship. A painting comes to represent something through a series of approximations that always introduce deviation at the same time as resemblance; the meaning of a painting includes the endless accidents that determine it as something other than what it represents. A digital image, on the other hand, is constructed from rules that delimit and specify its relationship to the represented, pinning down its meaning according to the model it emulates. Whereas the actual world, including the representations in it, immerses everything in a rich field of infinite ramification, guaranteeing that meanings will always be produced anew, the digital world instead restricts meaning to the predeterminations of the rules that govern its operations.

Representation's unique status in the digital derives from the digital's essential reliance on code. Bits become meaningful only when that meaning is assigned from without as a code according to a logic of representation. The parts of a digital object, the pixels of a digital image, the submillisecond samples that, in sequence, constitute a digital sound—each is determined not by its material instantiation as a bit (or sequence of bits), which bears no inherent relation to representation or to any represented, but by a set of codes that associate bit values with piecemeal representations. As discussed in chapter 4, codes regulate where a bit sequence begins and ends, how numbers are associated with pixels to determine color, and how to interpret a sequence of bits as voltage values that can induce shifting magnetic fields that cause oscillations in a speaker driver that make the air vibrate to generate sounds.

Inasmuch as any digital representation must be encoded in bits, there is necessarily a prior decision about how that representation will be constructed. There is a formal relation that governs the association between a digital image and the object or objects that it represents. The data in a database represent events or people or texts or corporations only because it has already been decided how certain sequences of bits will be interpreted as a code that decodes to yield those data. Thus, representations in the digital bear stricter relationships to their representeds. A digital representation is formed by a deliberate deployment of codes, such that a formula oversees the relationship to the represented. In actual representation, there is no bottom line; the representational relation must invent itself each time, so that it is never the same thing that makes an image an image of what it represents. Each photograph, each painting, each utterance represents by a new relation every time. But in the digital, representation is mostly though not entirely a matter of a logic, a representational logic, that governs what form is represented, and thus how it represents. The representation invariably exceeds its form, as any representation must. But this excess is less significant in digital representation, which does not invent its representational relationship, but instead appeals to a prescribed code, a formula that overcodes digital representation and reduces the excess that invites the production of new meaning.

According to the ontology of the actual, *everything* can change, everything is variable, even death, certainly taxes. In the digital ontology, there are things that can change, but also things that stay the same. This distinction between dynamic and static is pervasive in the digital, characterizing different aspects of the digital device, not just software versus hardware, but RAM and ROM, variable and constant, quality and part, content and form. Digital objects and digital processes have fixed features and variable features, a division ingrained in the ontology of the digital. Beyond the embedded division of hardware and software, there are versions of the dynamic–static dichotomy within the software, readable from the source code. The code might, as in the example presented in chapter 4, institute a loop in which an action is carried out iteratively, each cycle through the loop bringing about a change in some data structure, or adding a new item to a list, or subtracting one from a counter. Here the change is in the counter, or structure, or list, but notably not in the loop itself, which stays the same even as it causes other things to change. The structure of the digital as always including both fixed points and parts in (logical) motion corresponds to the necessity of an outermost frame, a bottom line, an ultimate, stable formula that decides what can vary and how.

Digital stasis does not imply a timeless permanence. At a different temporal scale, even the fixed elements of the digital morph and iterate, rebalancing stasis and dynamism through the progress of digital history. Restart the machine, update the software, install additional memory, upgrade the operating system, buy a new computer, move to the cloud, get the latest gadget, adopt a new frame. Each of these maneuvers alters what had formerly been fixed, establishing new stases and pulling back to open a new perspective. In this sense, time seems like the digital's only real contact with contingency. How will the industry develop? What pursuit will next undergo digitization? Which features will matter in the new application? When will that bug get fixed? Nothing is wholly firm, for the digital as historical progression is, like all history, exposed to contingency and never entirely predictable.

Inside the digital device, a clock tracks the rapid beats of the machinic heart, marking the stuttering temporality of the digital process that moves from stasis to stasis, from one discrete state of the machine to the next, where each state specifies the binary values of every bit, a huge sequence of 0s and 1s, any of which might change at the next tick of the clock. But other scales of time include the digital not as a sequence of successive states, but as a continuous history: the era of the personal computer, the rise of Web 2.0, the time since last restart, how long you've been playing that game of solitaire, or just clicking the mouse, saving a file, choosing a tool from the toolbar, writing another sentence. These are all historical events, momentary (or lengthy) distractions of the digital device, each of which punctures and then restores the hermetic world of the digital. Every interaction with the machine reshapes its forms to fit the world anew. The digital does not thereby relinquish its necessity, does not soften its edges or somehow succumb to accident. Its response to interaction remains resolutely digital: the production of form and only form. This production does not directly encounter contingency, but traces its shadow, what remains of the new once it settles into stillness. The digital carves out in its codes those structures left in the wake as the wave of contingency sweeps across the earth. The actual, in its vibrancy, its liveliness, ensures that the digital too will have new tasks to perform, new codes to honor, new forms to generate. Thus is there always the next update, a reconfiguration of bits related through time, by value and place, to act out a key-framed dynamism, shifting forms that mirror the structures of a world in which contingency runs rampant but leaves the outlines of objects and events in the ruins strewn behind it.

This appeal to time as the digital's closest brush with contingency is also Luciana Parisi's strongest claim. She characterizes the algorithm as a "new kind of model, which derives its rules from contingencies and open-ended solutions" (2013, 2). In defense of this view, she invokes Gregory Chaitin's number, omega, and the undecidability that makes that number so curious, but this is an overreach, an attempt to locate a contingency in the algorithm itself, in the operation of the digital. The bulk of her argument instead recognizes that the real action is in the vast spread of digital representation as it meets always new demands and constructs new forms in response to data that had not been anticipated. Parisi observes the digital as it occupies more terrain—organic, monstrous, insatiable—expanding at its edges to make a contact always with its outside. It does not thereby absorb contingency, but it gnaws at the actual, and we have been feeding it plenty.

Less abstractly, time is also the medium in which to understand the meaning of *another rule,* as in "One can always add another rule." The addition of another rule cannot imbue the digital with a contingency, but it does respond to the contingency of the actual.

We can increase resolution, take more exceptional cases into account, make the formula more sensitive to its context, program a coordination between the lemon's simulated texture and its simulated degree of ripeness, always by way of more code or more data. It is a matter of squeezing the contingent, the unnecessary, into the bounds of necessity, taking some part of the world and turning it into a rule, to stake out a more rigid actuality but a more supple or capacious digital system. To do so, to grasp the world as parts and qualities, as an aggregate of atoms, as a strict space of rules, is to squeeze out the contingency that engenders that world. But at the meeting of the digital and the actual, where the digital selects pieces of the world by a mimesis of pure form, something accidental, something not already necessitated is confronted. The digital faces contingency as a beyond, as what is not yet tamed, and this beyond is essential to digital operation, for it gives the digital its possibility of making meaning. This outside that threatens the digital with meaninglessness is the only thing that demonstrates to it its pure formality.

Contingency cannot be included in the digital, but it is nevertheless essential to it, for only in relation to a productive world, a world in which meaning is generated, does the digital also become meaningful. The digital accrues meaning through a dynamism of the representational logic, its shifting meanings arising as it is perpetually repurposed to address more of our world and more of our lives. The shifts over time, the encounter between the lively actual and the formal digital, happen continuously, whenever the machine is in use. But it is a question of intensity, of how creatively the actual challenges the digital to recode, to deploy itself anew. Does this point to an ethics of digital engagement, a challenge to us, the users, that we challenge the digital in turn?

The digital kills what it eats, renders its captives lifeless in order to keep them captive. It strips its prey of flesh, tossing aside the meat and keeping the bones to play games of memento mori. But this should not sound sinister, for the digital is only a tool, and its reductiveness, its rejection of contingency, founds its most powerful capacity. It is up to us to use the digital responsibly, to resist its alluring ideology that would train us to see the world, even apart from digital technology, as posits subject to a rational order and so fully instrumentalizable. When we use the digital habitually, when we use it thoughtlessly, artlessly, without carving out a new form in a new place, then we become uncreative and adopt its ideology. The digital must be exposed to contingency in the only way that it can be, in a relation to the world that can make something different even out of what is always the same. (The same difference, no matter how many rules one adds.) Employ the digital as a challenge, take it where it does not fit, and never take its procedures for granted. With vigilance, we can both avail ourselves of its fantastic value and ward off its delirious invitation to think like a machine.

Acknowledgments

As against an ongoing generational shift in humanities research, writing is usually for me a solitary endeavor. Fearful, perhaps, that others are more expert or just smarter than I am, I choose to write about ideas outside established disciplinary categories. Such an approach almost guarantees originality, since few others are addressing the questions I am investigating, but by the same token it risks irrelevance: if the topic were worthy of investigation, wouldn't people already be writing about it?

By that standard, this book is at least a small departure (improvement?) from my usual writing habits. Having secured tenure some years ago, I have felt more at home in my academic appointment at Dartmouth College and have been freer to enjoy the collegiality and intellectual companionship offered in this isolated liberal-artsy town. I have been more willing to share my interests with my trusted peers, and have received invaluable counsel from them, both much-needed moral support and incisive intellectual direction. As such, there is little meaningful distinction in this case between friends and colleagues, and the people I mention here have made this book possible by playing both roles.

Andrew McCann demonstrates repeatedly that it is possible to honor the highest ethical commitments and fight the good fight even knowing that it is thankless and Sisyphean work. His faith in me has held me steady through many bumps and narcissistic injuries.

George Edmondson and Michael Chaney have, each in his way,

guided me back to the real world at moments when my perspective became warped. I strive to repay their generosity in kind. They can't possibly know how much they have helped me.

Nirvana Tanoukhi has been a breath of fresh air and a constant inspiration. She leads the fight for the freedom to think without compromise, and her forthright manner, selfless curiosity, and remarkable intelligence leave me humbled that she has chosen me as an interlocutor, friend, and fellow thinker.

I have been fortunate to encounter others in this small community who sympathize with my critical regard for the digital and advance that critique in their own work. Jed Dobson offered thorough, detailed, and extremely helpful comments on a draft of this manuscript and is one of very few people I know with the training and insight to move seamlessly between philosophical inquiry and deep technical know-how; if he did not assure me otherwise, I would feel, in comparison to him, ill equipped to write this book. Daniel Affsprung's master's thesis and careful commentary on early drafts of this monograph were valuable in many ways, but foremost because they showed me that my intuitions were not wholly misguided, as they are shared by one so thoughtful and informed.

Kyle Booten has given so much of himself: encouragement, warmth, and more than one bottle of fine liquor. I suspect we both know who was the mentor in this relationship. I look forward to reading more of his great work.

Beyond the haven of Hanover, New Hampshire, three scholars of the digital played a determinative role in the development of this book. Rita Raley, Alex Galloway, and Matt Fuller participated in a review of a draft of the book, and their enthusiastic engagement and generosity gave me the resolve to heed their astute and careful criticisms, improving this book dramatically. That Alex has also authored a fantastic foreword is more than I could have asked.

I offer my immense gratitude to the one anonymous reader for the University of Minnesota Press who read carefully, grasped the aim and import of the project, and offered piercing but supportive criticisms, which also improved this book considerably. This fantastic experience will lead me to regard all of my colleagues in the Academy with a bit more openness and a bit less cynicism.

I cannot adequately acknowledge the indispensable place of Sally Ackerman in my work and my life; her influence is incalculable. She is my first reader, my sounding board, and the soil in which my thoughts grow. She can justly be called an author of this work, for she has nurtured me, and nurtured it, for a very long time.

Notes

Introduction

1. Undertaking a technophilosophical approach distinct from canonical digital studies, this book would not exist without those foundational works of the past few decades that have established the digital as an analytic object and evinced so many consequences and possibilities of digital technologies. I list here some of the books in digital studies that have been important, and especially important for me, but are not otherwise cited in this book, for their influence herein is diffuse. Some of these works practice the cultural theoretic analysis prevalent in digital studies, but these titles are notable because they all also step back from the particular and explore the nature of the digital in general, of digital ontology, epistemology, ethics, or aesthetics. The size and diversity of the field, which has grown in proportion to the spread of digital technology, ensures that this list elides much essential scholarship: Lev Manovich, *The Language of New Media*; Mark Hansen, *New Philosophy for New Media*; N. Katherine Hayles, *How We Became Posthuman, Writing Machines,* and *My Mother Was a Computer*; Ian Bogost, *Unit Operations: An Approach to Videogame Criticism*; Friedrich Kittler, *Literature, Media, Information Systems*; Wendy Hui Kyong Chun, *Control and Freedom, Programmed Visions: Software and Memory* and the coedited collection *New Media, Old Media: A History and Theory Reader*; Sherry Turkle, *The Second Self, Life on the Screen,* and *Alone Together*; Alexander Galloway, *Protocol: How Control Exists after Decentralization* and *The Interface Effect*; Pierre Lévy, *Becoming Virtual: Reality in the Digital Age*; Jay David Bolter and Richard Grusin, *Remediation: Understanding New Media*;

Hubert Dreyfus, *What Computers Can't Do*; Matthew Kirschenbaum, *Mechanisms: New Media and the Forensic Imagination*; Michael Heim, *The Metaphysics of Virtual Reality*; Adrian MacKenzie, *Transductions: Bodies and Machines at Speed*; Johanna Drucker, *Graphesis: Visual Forms of Knowledge Production*.

1. Approaching the Digital

1. Humanist scholarship on the digital is not to be confused with Digital Humanities. Though Digital Humanities has redefined itself more than once over the two decades or so of its existence, and has recently embraced certain critical examinations of the digital, its core practice is not the study of the digital but rather the employment of digital, computational tools to advance humanist scholarship. The present book, along with many others, offers a complementary approach, applying humanist critique to the digital rather than applying digital tools to humanist researches.

2. What Does the Digital Do?

1. The term *worldview* and the phrase *way of seeing* both misconstrue a crucial nuance of Heidegger's presentation, and not only because of their bias toward the visual (a bias that Heidegger sometimes shares). For Heidegger, *Ge-stell* refers both to the frame that we erect so that we encounter the world in a particular way (as standing-reserve) and also to the frame around *us* that reveals the world exclusively as *Ge-stell*. Heidegger hesitates between an understanding wherein we enframe the world and the converse structure wherein the world enframes us. By contrast, the notion of a *worldview* points in only one direction, from the subject outward, and lacks the productive ambiguity between person and world that is the signature gesture of Heidegger's philosophy.

2. Digital technologies rely pretty much exclusively on the binary code, but it should be noted for the sake of accuracy that it is discreteness and not binarity that provides the real principle of operation. One could build digital machines around a three-valued code or a thirty-valued code, and they would be able to accomplish the same things that current digital technologies do with a binary code. The de facto standard is binarity for a number of historical reasons, including the ready availability of materials that can exhibit binarity: an electrical

circuit can easily represent two states, for example, one in which electricity is flowing and another in which electricity stops flowing, and one could use many such circuits to create a machine that employs a binary code to do its work. In fact, this is more or less how computers work. Binarity also represents a certain minimum, the smallest distinction that can represent all distinction, so that there is something mathematically tidy or parsimonious about it. (In his 1703 essay "Explanation of the Binary Arithmetic," G. W. Leibniz affirms the value of this parsimony and hails the binary as "the most fundamental way of reckoning for science," a reduction of numbers to their "simplest principles.") Mathematics usually prefers the elegance of simpler expressions and minimal bases to express more complicated systems.

3. The opposite claim is also popular: digital machines are anything but neutral and actually carry with them all the biases of the cultures that built them. Such criticisms—in many cases they are called *algorithmic* criticisms—typically locate the source of the bias not in the digital execution of the algorithm (the digital carries out the orders it is given), but in its extradigital design. That design might include sets of data that are not properly representative of the population under consideration, a problem all too likely because correlated variables can hide a biased data set behind a presumptively neutral quality. For example, in the United States, the alarmingly disproportionate number of African Americans in prison means that a study of ex-convicts that chooses a random selection of former prisoners as its test population will likely reflect that disproportionately high African American representation, such that statistically significant conclusions drawn on the basis of that test population, which is intended to indicate something about former prisoners, might really be capturing a quality more common to African Americans generally. In any case, the digital has no leeway to deviate from the encoded instructions it is given, such that any error or bias must have been put in those instructions; the digital cannot originate bias, except an absolute bias toward the positivism that is its basis of operation.

4. Computer-science pioneer Fred Brooks describes this irreducible labor in *The Mythical Man-Month,* where he argues that the labor of programming is not so much about coaxing a recalcitrant machine into doing one's bidding as it is about matching the complexity of the task to be accomplished with a similarly complex logical-digital structure.

3. Ontology and Contingency

1. In his influential chapter "On the Superiority of the Analog," Brian
   Massumi anticipates many of this current book's critical analyses of
   the digital. Particularly relevant to the analysis herein is the distinction he articulates between possibility and potential. Whereas
   potential is the indefinite not-yet that this book identifies with contingency, possibility is the availability of a set of predefined conditions that could become actual, which neatly describes the operation
   of the digital: "Digital technologies in fact have a remarkably weak
   connection to the virtual, by virtue of the enormous power of their
   systematization of the possible" (137).
2. How to interpret the data of quantum mechanics remains an unresolved matter. There are theories that reestablish an individual determinism at the quantum level, but they have not earned a consensus
   accord.
3. Two hundred fifty years before Meillassoux, David Hume already
   concluded that we have no choice but to assume that the future
   will be more or less like the past, even though we know of no good
   reason to support this belief. Meillassoux's contribution shows that
   there can *never* be reason to believe this, because it is in fact an ill-
   founded belief.
4. Though this description will later be contrasted with the ontology of
   the digital, it is notable that this ontology accords well with the description of the digital object offered by Yuk Hui (2016). According
   to Hui, a digital object, like any object, is sustained in its objecthood
   by its internal interdependencies, as well as another set of relations
   between those internal relations and various externalities. His more
   recent monograph on recursion situates digital objects among objects and events more broadly.
5. Fractal-based computer simulation of natural phenomena is a
   common technique in computer graphics and some other modeling
   applications. Not only does this technique approximate real natural
   forms, but it makes very efficient use of computing power: a complex form can be algorithmically specified just once but then applied
   at multiple scales, which gives the appearance of greater complexity
   and disguises, at least to some extent, the rule-bound character of
   the generating algorithm. We might think of fractal techniques in
   computation as an attempt to simulate contingency, though notably, such techniques tend to have a certain recognizable aesthetic;

once one has seen enough of them, they reveal their ruliness in their appearance.

4. Ontology of the Digital

1. The transparency of these simple table look-ups does not imply the transparency or overall simplicity of the successive calculation of billions of bitwise operations. That is, one cannot in general know what the result will be of a vast number of successive table look-ups in which the inputs to one table are the outputs of previous calculations. In fact, early in the computer age, Alan Turing proved that there is no general mechanism that can always determine, for any sequence of calculations, whether that sequence will eventually arrive at a concluding result or, on the contrary, will continue calculating forever without ever arriving at a terminal state.

2. Technically, the CD standard does use successive dots on the disc's surface to represent bits, but instead of assigning 0 or 1 to each dot based on shininess, the binary encoding registers the *change* from a shiny dot to its unshiny neighbor (or vice versa) as a 1 bit and the nonchange from a shiny bit to its shiny neighbor (or unshiny to unshiny) as a 0. This seemingly unnecessary complication of encoding makes the resulting storage system more robust, less likely to exhibit certain kinds of physical errors. Also, technically, the relevant distinction isn't between a shiny and an unshiny spot on the disc's surface, but rather whether the shininess is right on the surface of the disc or slightly recessed, such that a laser light reflecting back from that surface must travel a little bit farther, hence take a little bit longer, before triggering the optical sensor waiting for it. The difference on a CD between a 0 and a 1, therefore, is measured by a clock that starts when the laser light is turned on and stops when it bounces back to the optical sensor. Or, rather, it is measured by comparing successive time measurements as the laser shines first on one spot and then on the next spot. Again, this seeming complication reduces the medium's susceptibility to certain kinds of reading errors, to make the system more robust.

3. Media theorist Friedrich Kittler draws out the consequences of this externalized ideality in the opening paragraphs of his 1995 essay "There Is No Software." He observes that, by relocating this act of abstraction from our human minds into digital machines, we have effectively ceded the very idea of reading (and writing) to those

machines: "The last historical act of writing may well have been the moment when, in the early seventies, Intel engineers laid out some dozen square meters of blueprint paper (64 square meters, in the case of the later 8086) in order to design the hardware architecture of their first integrated microprocessor."

4. Programmers who wish to direct the machine at the granular level of the CPU usually work not by reading and writing sequences of 0s and 1s, but instead by using human-readable alphanumeric abbreviations for those bit sequences, such as MOVE or INCR. Those abbreviations, which correspond one-to-one with machine language commands, are called "assembly language." A program written in alphanumeric assembly language must be translated into the corresponding bit sequences in order to execute on the machine.

5. A computer chip has a bus size, which refers to the number of bits that it can accept simultaneously as a single input. A number of tiny metal legs sometimes give microchips a spider-like appearance, and those legs attach to nodes on the circuit board to accept input and issue output; the number of legs typically indicates the bus size, the number of bits for a single instruction that enters the chip all at once.

6. This description of a CPU executing relatively simple commands applies most readily to the popular class of CPUs that employ a reduced instruction set computing architecture (RISC). There are also complex instruction set computers (CISC), chips that generally execute more processing steps per command. In either case, the steps themselves remain fairly straightforward, as in the examples given in the text.

7. To generate a number that is close enough to truly random, a digital machine starts by grabbing a value from some part of the machine that varies rapidly, such as the digits of the system clock that counts millionths of a second. Then that seed value is run through some wacky, nonlinear arithmetic that has the effect of decorrelating values, so that two numbers that are close together will no longer necessarily be close when each has been run through the wacky arithmetic. The arithmetic is a deterministic function, but complicated enough that its pattern is not easily discernible.

8. The foundational claim of Johan Huizinga's famous cultural theory of play, *Homo Ludens,* is that play is an irreducible and essential animal activity, especially well developed in humans. He too identifies play with a kind of freedom, but also recognizes that part of the pleasure lies in deliberately restricting that freedom through agreed rules. The space in which the rules are held firm, in which the game takes

place, constitutes a miniature world unto itself, which Huizinga dubs "the magic circle." Though some games, such as chess, are wholly determined by their rules, which define and delimit the possible moves in the game, other games, such as charades, establish a space of restricted possibility that still gives creative freedom room to roam.

9. It is possible to construct the equivalent of all sixteen logic gates starting with only one of them, though the starting gate would have to be either NAND or NOR. This construction is left as an exercise for the reader! Some gates are easier or more efficient to implement in hardware than others, so most chips are constructed with some subset of the sixteen available gates, implementing the missing gates as needed by combining gates in that subset.

10. This image of a rational indexing scheme to identify any bit within a digital machine is a simplification, but not an obfuscation. In practice, when programming, or even when a program is running, bits are typically referenced by indirect indexing, where a label or number refers to a place in the machine's memory that stores an index to another place in the machine's memory where the bits of interest are located. Indirect indexing sacrifices a modicum of efficiency—it takes additional clock cycles to retrieve the index and *then* retrieve the data in question—but adds a layer of abstraction that makes it easier to move a program (or operating system) from one machine to another, even if those machines have different memory architectures, and thus different native indexing schemes. Further, the numerical indexing scheme in a digital machine usually refers not to individual bits, but to a group of eight bits in sequence known as a *byte*. Accessing individual bits within a byte requires additional logic.

5. From Bits to the Interface

1. Quantum computing appears to violate the strict independence of bits, as it relies on quantum entanglement, wherein bits are locked in relations of mutual determination through physical operations that tie them together. But, even in that case, qubits (as they are called in quantum computing) are initially independent of each other and become entangled only through procedures that act on groups of qubits together. Analogously, an algorithm might impose a relationship among bits, including code that checks whether a given bit has changed and, if so, updates related bits in response to that change. But any relationship among bits would be a contrivance; it would

require deliberate enforcement of that relationship in code, because a bit never has any inherent relationship to another bit.

2. The principle invoked here is strictly correct: the digital machine can do nothing that it is not explicitly programmed to do. No action, no input is inherently meaningful in the digital, but takes on assigned meaning only through programmed behavior. However, it is not strictly true that each software engineer has to do all of this coding herself; instead, most significant software packages are written in part by borrowing and incorporating prewritten code libraries that include algorithms that do a lot of the most frequently needed processing. A code library might have specialized algorithms for dealing with three-dimensional spaces or for displaying text on the screen or for performing high-level mathematical calculations, so that the programmer can simply invoke those algorithms when necessary rather than writing that code herself. Moreover, the operating system of a computing machine tends to have a lot of basic functions already coded in it. Thus, personal computers have software built in to the operating system that detects mouse clicks; a programmer need only ensure that her code responds to mouse-click messages sent by the operating system, but she needn't write the click-detection code herself.

3. At the end of his elliptical, hyperbolic, and beautiful 1995 essay "There Is No Software," Friedrich Kittler proposes that only a computing device essentially indistinguishable from material reality could match the computational power of the universe itself. Any device that relies on a code of programmable discrete elements with exact definition, thereby excluding the noise of limited precision, will necessarily fall well short of the chaos of the real.

4. The broadest and most universal measure of a batter's success is a batting average, which is the ratio of the batter's number of hits to times at bat, but the example here considers an individual pitch, not an at-bat.

5. Especially in the age of "moneyball" baseball, where many aspects of the professional game are managed or strategized using data analysis, pitches are often chosen according to statistical analyses of a hitter's record. While the statistical results may be treated as rules for the purposes of tactical decisions, they exert no influence on the hitter's future performance, but constitute a best guess on the often-validated assumption that the future will be more or less like the past. Moneyball exemplifies the way in which digital data collection and analysis has pushed aside contingency, such that the

world of professional athletics is subjugated to its statistics rather than dominated by the contingency that makes sports so compelling for players and spectators. On the other hand, it is only because contingency reasserts itself at the edges of any game and especially at crucial moments of play, only because *we know that we don't know* what is going to happen, that sports remain such a captivating spectatorial activity.

6. Though most well-known, the Turing test is not always regarded as the best criterion of artificial intelligence. In Alan Turing's paper describing this test, he regards it more as one interesting experiment, one threshold to consider, rather than the sine qua non of intelligence criteria.

7. There are techniques of software engineering that make use of evolutionary principles. Neural networks evolve their own rules, as discussed above. But *evolutionary software development* models bioevolution even more closely: one begins with one or more algorithms, then induces random variations in the code of those algorithms and observes the resulting algorithmic behavior. When a variant algorithm does something interesting or helpful or unexpected, it is preserved as a member of the next generation, and those surviving algorithms are allowed to mutate again. It is even conceivable to let two surviving algorithms share code with each other, to simulate a kind of sexual reproduction that may make for more robust evolution. Notably, though, even this evolutionary software development technique still relies on part-by-part evolution; any new behaviors result from specific alterations to the code that are, in effect, substitutions of new rules for old rules. Unlike the expansive and contingent nuance of real evolution, where everything inside and outside of the individual connects to everything else, digital evolution produces only part-to-part relations.

8. Microchips and other microsensitive materials are generally manufactured in "clean rooms," where dust particles have been swept from the air, thus minimizing the likelihood of faulty manufacture due to dust. The most common—but still extremely uncommon— failure of bitwise calculations in microchips (or chip-based memory) occurs due to the occasional particle of beta radiation striking the chip's surface, which can cause an incorrect read of the bit at that location. It is at least curious that beta radiation scrambles the codes of computers but also the code of life, causing cancerous mutations in the genetic code stored in each living cell.

6. What Does the Digital Do to Us?

1. There are interfaces that respond to a raised palm and other hand motions, usually using a camera to track the user's gestures. Whatever can be discretized, rendered as a determinate and distinct pattern, can become a recognized gesture in the digital. The possibilities are endless, so long as one has instructed the digital device regarding exactly what to look for.

2. A Ouija board calls for multiple hands of multiple people, which helps immensely to demote the will of any individual. Even a deliberate exercise of will would likely be thwarted by the combined resistance of the others, which, moreover, changes unpredictably. In 2014, the video-streaming service Twitch hosted a series of events in which every viewer could send commands simultaneously to a shared game of *Pokémon,* such that those eight thousand or so commands tended to undo or overdo each other, making for very labored and clumsy gameplay. As on the Ouija board, the individual wills of the many simultaneous *Pokémon* players generated a disunified chaos of competing desire that only haltingly approached any goal, though the game was eventually successfully completed.

3. In fact, the Game of Life is a Turing complete system, meaning that it can reproduce the activity of any digital computer, or that it can run any program. Given a big enough matrix of cells, it can therefore produce as much logical complexity as any digital device.

Bibliography

Adorno, Theodor, and Max Horkheimer. (1947) 2002. *The Dialectic of Enlightenment: Philosophical Fragments*. Translated by Edmund Jephcott. Stanford, Calif.: Stanford University Press.

Bogost, Ian. 2008. *Unit Operations: An Approach to Videogame Criticism*. Cambridge, Mass.: MIT Press.

Bolter, Jay David, and Richard Grusin. 2000. *Remediation: Understanding New Media*. Revised edition. Cambridge, Mass.: MIT Press.

Brooks, Frederick P. 1987. "No Silver Bullet: Essence and Accidents of Software Engineering." *Information Processing 86: Proceedings of the IFIP 10th World Computer Congress, Dublin, Ireland, September 1–5, 1986*, 1069–76. IFIP Congress Series. New York: North-Holland.

Brooks, Frederick P. 1995. *The Mythical Man-Month: Essays on Software Engineering*. Anniversary ed. Reading, Mass.: Addison-Wesley.

Campbell, Jim. 2000. "Delusions of Dialogue: Control and Choice in Interactive Art." *Leonardo* 33 (2): 133–36.

Cayley, John. 2018. *Grammalepsy: Essays on Digital Language Art*. New York: Bloomsbury Academic.

CD Projekt Red. 2020. *Cyberpunk 2077*. Sony Playstation 4 ed.

Chaitin, Gregory. 2011. "How real are real numbers?" *Manuscrito* 34 (1): 115–41. doi.org/10.1590/S0100-60452011000100006.

Chun, Wendy Hui Kyong. 2008. *Control and Freedom: Power and Paranoia in the Age of Fiber Optics*. Illustrated edition. Cambridge, Mass.: MIT Press.

Chun, Wendy Hui Kyong. 2013. *Programmed Visions: Software and Memory*. Reprint edition. Cambridge, Mass.: MIT Press.

Chun, Wendy Hui Kyong, and Thomas Keenan, eds. 2005. *New Media, Old Media: A History and Theory Reader*. New York: Routledge.

Del Santo, Flavio, and Nicolas Gisin. 2019. "Physics without Determinism: Alternative Interpretations of Classical Physics." *Physical Review A* 100 (6): 062107.

Deleuze, Gilles. 1990. *Logic of Sense.* Translated by M. Lester and C. Stivale. New York: Columbia University Press.

Deleuze, Gilles. 1992. "Postscript on the Societies of Control." *October* 59: 3–7.

Derrida, Jacques. 1988. "The Purveyor of Truth." Translated by Alan Bass. In *The Purloined Poe: Lacan, Derrida, and Psychoanalytic Reading,* edited by John Muller and William Richardson, 173–212. Baltimore, Md.: Johns Hopkins University Press.

Douglas, Christopher. 2002. "'You Have Unleashed a Horde of Barbarians!': Fighting Indians, Playing Games, Forming Disciplines." *Postmodern Culture* 13 (1). pmc.iath.virginia.edu/issue.902/13.1douglas .html.

Dreyfus, Hubert L. 1986. *What Computers Can't Do: The Limits of Artificial Intelligence.* Revised ed. 7th print. Perennial Library. New York: Harper & Row.

Dreyfus, Hubert L. 2009. *On the Internet,* 2nd revised ed. New York: Routledge.

Drucker, Johanna. 2001. "Digital Ontologies: The Ideality of Form in/and Code Storage—or—Can Graphesis Challenge Mathesis?" *Leonardo* 34 (2): 141–45.

Drucker, Johanna. 2014. *Graphesis: Visual Forms of Knowledge Production.* Cambridge, Mass.: Harvard University Press.

Dunn, David. 2002. "Nature, Sound Art and the Sacred." Art & Science Laboratory / Music Language, and Environment. davidddunn .com/~david/writings/terrnova.pdf. Accessed August 2020.

Fazi, M. Beatrice. 2018. *Contingent Computation: Abstraction, Experience, and Indeterminacy in Computational Aesthetics.* New York: Rowman & Littlefield.

Friedman, Ted. 1999. "*Civilization* and Its Discontents: Simulation, Subjectivity, and Space." In *On a Silver Platter: CD-ROMs and the Promises of a New Technology,* edited by Greg Smith, 132–50. New York: New York University Press.

Galloway, Alexander R. 2004. *Protocol: How Control Exists after Decentralization.* Cambridge, Mass.: MIT Press.

Galloway, Alexander R. 2006. *Gaming: Essays on Algorithmic Culture.* Minneapolis: University of Minnesota Press.

Galloway, Alexander R. 2012. *The Interface Effect.* Malden, Mass.: Polity.

Hansen, Mark B. N. 2004. *New Philosophy for New Media.* Cambridge, Mass.: MIT Press.

Hayles, N. Katherine. 1999. *How We Became Posthuman: Virtual Bodies in Cybernetics, Literature, and Informatics.* Chicago: University of Chicago Press.

Hayles, N. Katherine. 2002. *Writing Machines.* Cambridge, Mass.: MIT Press.

Hayles, N. Katherine. 2005. *My Mother Was a Computer: Digital Subjects and Literary Texts.* Chicago: University of Chicago Press.

Heaven, Will Douglas. 2020. "OpenAI's New Language Generator GPT-3 Is Shockingly Good—and Completely Mindless." *MIT Technology Review,* July 20, 2020.

Heidegger, Martin. 1962. *Being and Time.* Translated by J. Macquarrie and E. Robinson. New York: Harper & Row.

Heidegger, Martin. 1977. "The Question Concerning Technology." In *The Question Concerning Technology and Other Essays,* translated by William Lovitt, 3–35. New York: Harper & Row.

Heim, Michael. 1994. *The Metaphysics of Virtual Reality.* New York: Oxford University Press.

Hui, Yuk. 2016. *On the Existence of Digital Objects.* Minneapolis: University of Minnesota Press.

Hui, Yuk. 2019. *Recursivity and Contingency.* New York: Rowman & Littlefield.

Huizinga, Johan. (1938) 1971. *Homo Ludens: A Study of the Play Element in Culture.* Boston: Beacon.

Kant, Immanuel. 1998. *The Critique of Pure Reason.* Translated by Paul Guyer. Cambridge: Cambridge University Press.

Kirschenbaum, Matthew. 2012. *Mechanisms: New Media and the Forensic Imagination.* Cambridge, Mass.: MIT Press.

Kittler, Friedrich. 1995. "There Is No Software." *CTheory.* journals.uvic.ca/index.php/ctheory/article/view/14655/5522.

Kittler, Friedrich. 1997. *Literature, Media, Information Systems.* New York: Routledge.

Lanier, Jaron. 2011. *You Are Not a Gadget: A Manifesto.* London: Vintage.

Leibniz, Gottfried Wilhelm. (1703) 1962. "Explanation of the Binary Arithmetic." In *Die mathematische schriften von Gottfried Wilhelm Leibniz, vol. VII,* edited by C. I. Gerhard, 223–27. Translated by Lloyd Strickland at leibniz-translations.com/binary.htm.

Lévy, Pierre. 1998. *Becoming Virtual: Reality in the Digital Age.* New York: Plenum Trade.

MacKenzie, Adrian. 2006. *Transductions: Bodies and Machines at Speed.* New edition. New York: Continuum.

Mandelbrot, Benoit. 1982. *The Fractal Geometry of Nature.* New York: W. H. Freeman.

Manovich, Lev. 2002. *The Language of New Media.* Reprint edition. Cambridge, Mass.: MIT Press.

Massumi, Brian. 2002. *Parables for the Virtual: Movement, Affect, Sensation.* Durham, N.C.: Duke University Press.

Meillassoux, Quentin. 2008. *After Finitude: An Essay on the Necessity of Contingency.* Translated by Ray Brassier. New York: Continuum.

Nietzsche, Friedrich. 1989. "On Truth and Lying in an Extra-moral Sense." In *Friedrich Nietzsche on Rhetoric and Language,* edited and translated by Sander Gilman, Carole Blair, David Parent, 246–57. New York: Oxford University Press.

Pajkovic, Niko. 2021. "Algorithms and Taste-Making: Exposing the Netflix Recommender System's Operational Logics." *Convergence.* May 17, 2021. doi.org/10.1177/13548565211014464.

Parisi, Luciana. 2013. *Contagious Architecture: Computation, Aesthetics, and Space.* Cambridge, Mass.: MIT Press.

Parisi, Luciana. 2019. "Critical Computation: Digital Automata and General Artificial Thinking." *Theory, Culture & Society* 36 (2): 89–121.

Ramsay, Stephen. 2011. *Reading Machines: Toward an Algorithmic Criticism.* Urbana: University of Illinois Press.

Russell, Bertrand, and A. N. Whitehead. [1910] 2011. *Principia Mathematica.* Volume 1. San Bernadino, Calif.: Rough Draft.

Saussure, Ferdinand de. 1959. *Course in General Linguistics.* Translated by Wade Baskin. New York: Philosophical Library.

Shannon, Claude. [1949] 1971. *A Mathematical Theory of Communication.* Chicago: University of Illinois Press.

Stern, Joanna. 2007. "The Way It Woz: Steve Wozniak on All Things Apple" (interview). *Laptop: Mobile Solutions for Business & Life,* October 26, 2007. myoldmac.net/interviews/20071026_TheWayItWoz-SteveWozniak-onAllThingsApple.html.

Stevens, Wallace. 1971. "The Emperor of Ice-Cream." In *The Collected Poems of Wallace Stevens,* 64. New York: Alfred Knopf.

Turing, Alan. [1950] 1999. "Computing Machinery and Intelligence." In *Computer Media and Communication: A Reader,* edited by P. Mayer. 37–58. Oxford: Oxford University Press.

Turkle, Sherry. 1984. *The Second Self: Computers and the Human Spirit.* New York: Simon & Schuster.

Turkle, Sherry. 1997. *Life on the Screen: Identity in the Age of the Internet.* New York: Simon & Schuster.

Turkle, Sherry. 2017. *Alone Together.* 3rd edition. New York: Basic Books.

Von Neumann, John. 1963. "General and Logical Theory of Automata." In *Design of Computers, Theory of Automata, and Numerical Analysis.* Volume 5 of *Collected Works.* Oxford: Pergamon.

Wittgenstein, Ludwig. 1958. *Philosophical Investigations.* Translated by G. E. M. Anscombe. New York: Macmillan.

Wittgenstein, Ludwig. (1921) 1981. *Tractatus Logico-Philosophicus.* Translated by C. K. Ogden. New York: Routledge.

Index

abstraction, 86, 90, 92, 126, 128, 131, 149, 221, 223

accidents, 6, 13, 15, 19, 34, 51, 52, 57, 59, 69, 78, 88, 92, 139, 158, 169–70, 201; contingency of, 162; eliminating, 42; endless, 206; experiencing, 160, 161; happy, 102; historical, 56; openness to, 188

action: rationalism and, 48; rule of, 159

actual, 125, 131, 210; concrete infinity of, 138; contingency and, 88, 121; digital and, xvi, 6, 138; learning in, 169

Adorno, Theodor, 35

advertisements, 17, 49, 104, 155, 168; pop-up, 154

African Americans, number of imprisoned, 219n3

AI. *See* artificial intelligence

algorithm, 22, 24, 47, 69, 88, 93, 98, 112, 116, 117, 127, 132, 134, 135–36, 138, 140, 144, 157, 169, 176; big data and, 181; chess, 119, 120; context-dependent, 49; digital, 8, 23, 219n3; encoded, 26; ethics of, 21;

fecundity of, 156; importance of, 8; limits of, 145; predetermined, 16; shape-drawing, 136; triggering, 97

Altman, Sam, 193

ambiguity, xiv, 30, 36–39, 204; contingency and, 85; form and, 114–15; ontological, 201

American Standard Code for Information and Interchange (ASCII), 96, 108, 116, 123

analog, xv, 173; digital and, 5, 98

analysis, 12, 102, 105, 186; criteria for, 141; data, 146, 224n5; digital, 163–64, 220n1; principal component, 187; statistical, 183; step-by-step, 117; technological, 8

AND gate, 88, 110

antipositivism, 17, 62, 73, 84

Apple, 11, 50

Aristotle, xiii, 18

arithmetic, 22, 89, 193; binary, 23, 27; nonlinear, 222n7

art: accidental/automatic, 180; making, 78–79, 159; sound, 73–74

artificial intelligence (AI), xv, 5, 35, 68, 179, 195, 225n6

Electronic Mediations

(continued from page ii)

Aden Evens is associate professor of English and creative writing at Dartmouth College and author of *Sound Ideas: Music, Machines, and Experience* (Minnesota, 2005) and *Logic of the Digital.*

Alexander R. Galloway is professor of media, culture, and communication at New York University, Steinhardt. He is the author of *Gaming: Essays on Algorithmic Culture* (Minnesota, 2006), *The Exploit: A Theory of Networks* (Minnesota, 2007), *The Interface Effect,* and *Laruelle: Against the Digital* (Minnesota, 2014).

Printed and bound by CPI Group (UK) Ltd, Croydon, CR0 4YY

27/10/2024

14580398-0002